OHM'S LAW COMPLEX CIRCUITS

Exercise #1

Find the voltage and amperage for each resistor.

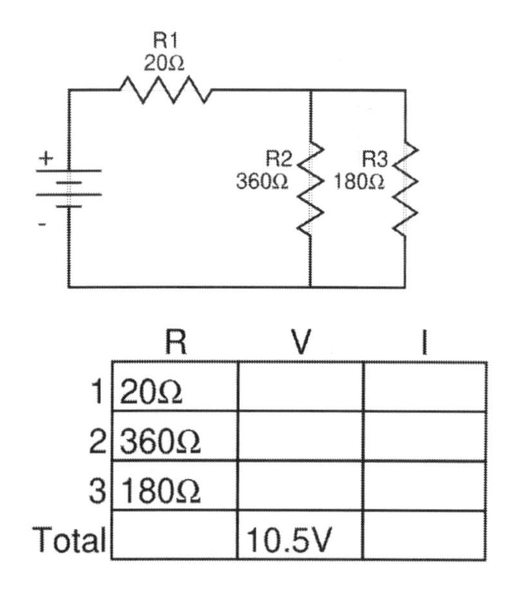

	R	V	I
1	20Ω		
2	360Ω		
3	180Ω		
Total		10.5V	

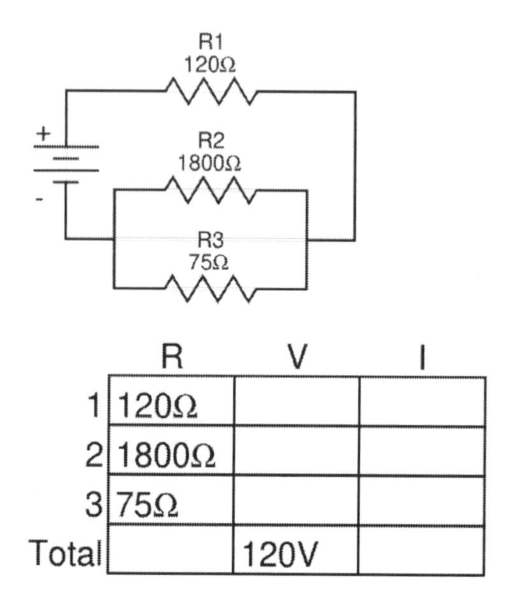

	R	V	I
1	120Ω		
2	1800Ω		
3	75Ω		
Total		120V	

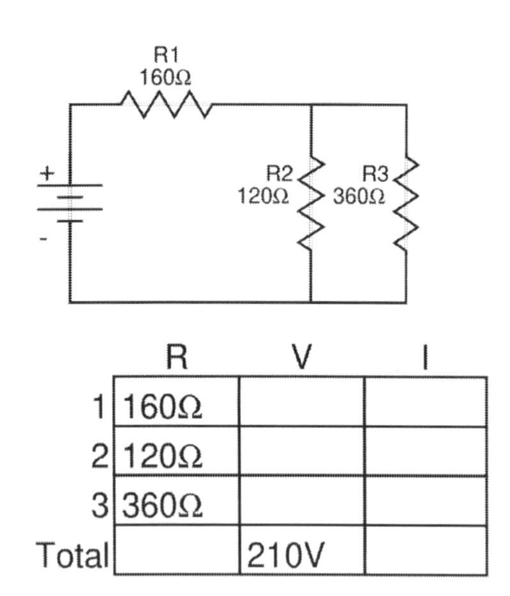

	R	V	I
1	160Ω		
2	120Ω		
3	360Ω		
Total		210V	

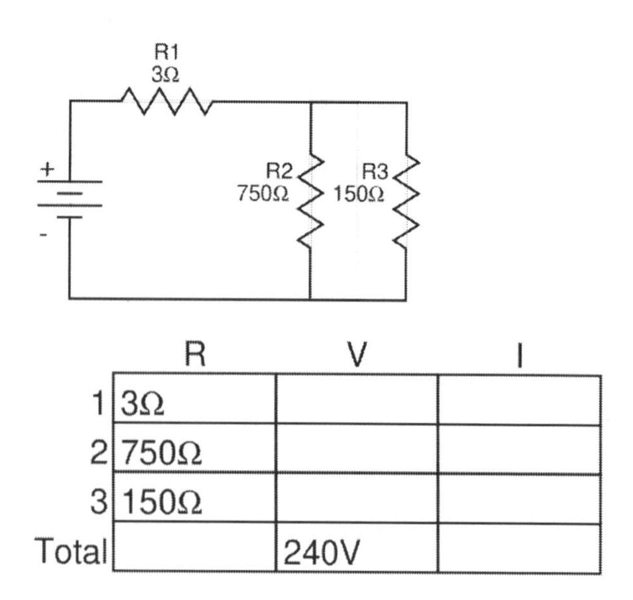

	R	V	I
1	3Ω		
2	750Ω		
3	150Ω		
Total		240V	

Exercise #2

Find the voltage and amperage for each resistor.

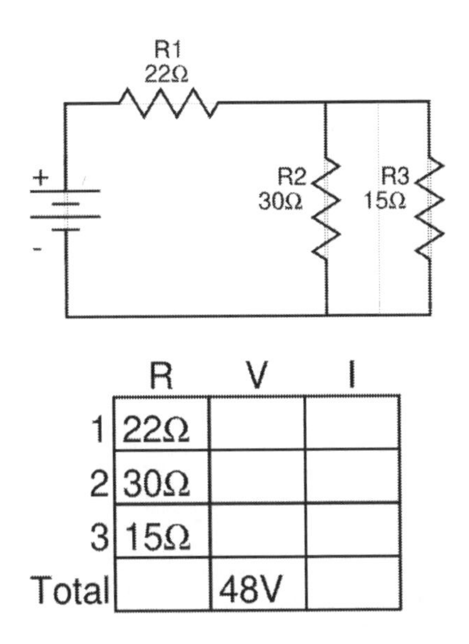

	R	V	I
1	22Ω		
2	30Ω		
3	15Ω		
Total		48V	

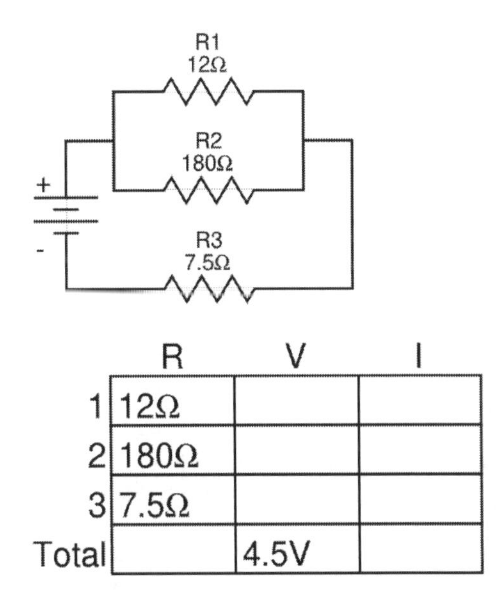

	R	V	I
1	12Ω		
2	180Ω		
3	7.5Ω		
Total		4.5V	

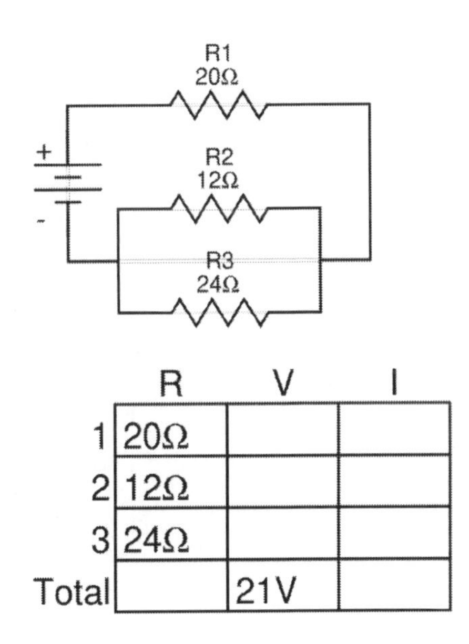

	R	V	I
1	20Ω		
2	12Ω		
3	24Ω		
Total		21V	

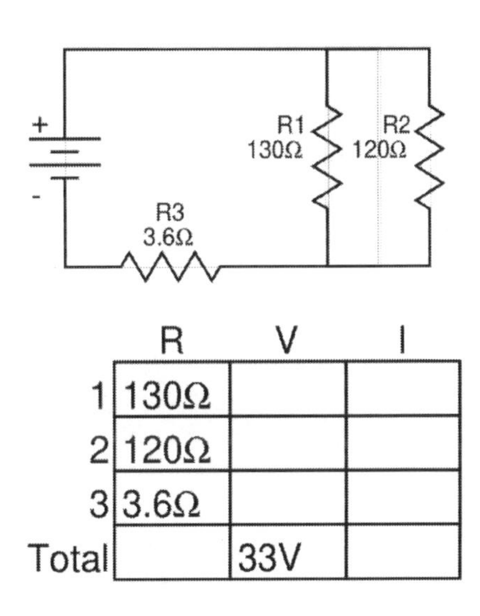

	R	V	I
1	130Ω		
2	120Ω		
3	3.6Ω		
Total		33V	

Exercise #3

Find the voltage and amperage for each resistor.

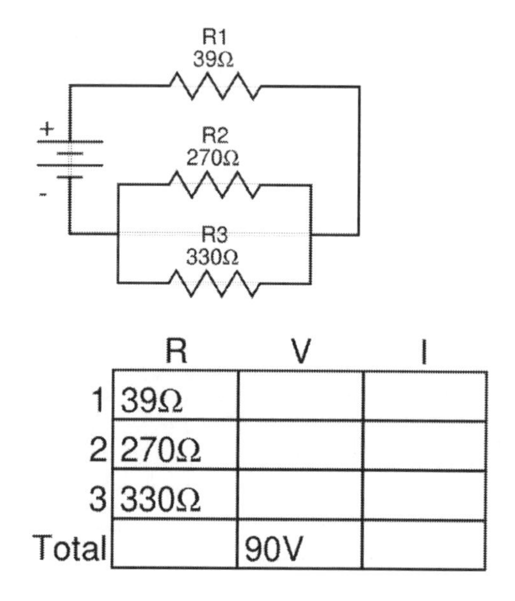

	R	V	I
1	39Ω		
2	270Ω		
3	330Ω		
Total		90V	

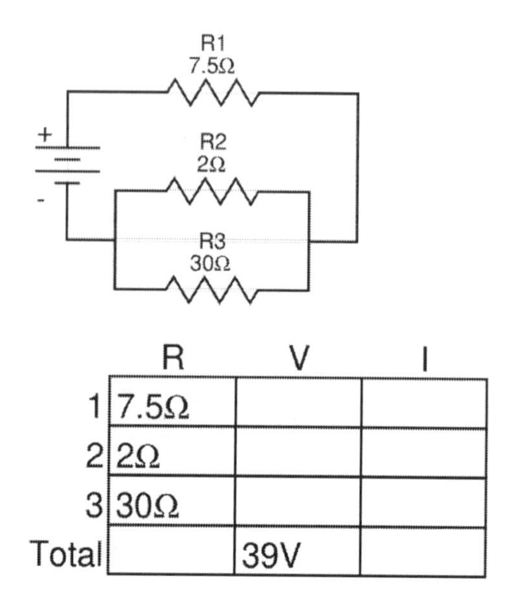

	R	V	I
1	7.5Ω		
2	2Ω		
3	30Ω		
Total		39V	

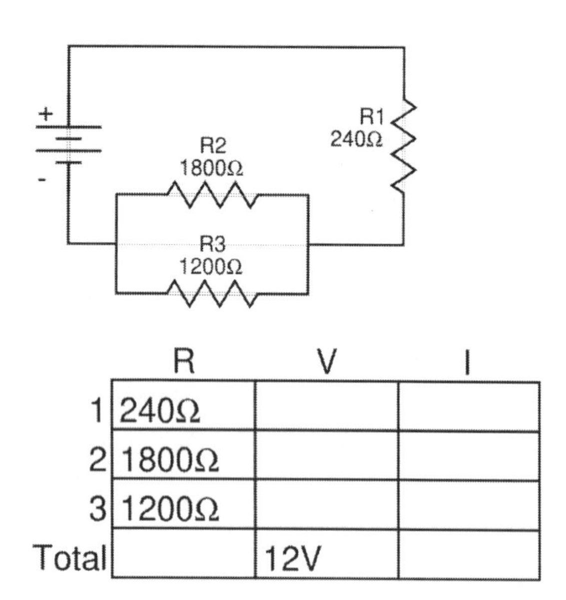

	R	V	I
1	240Ω		
2	1800Ω		
3	1200Ω		
Total		12V	

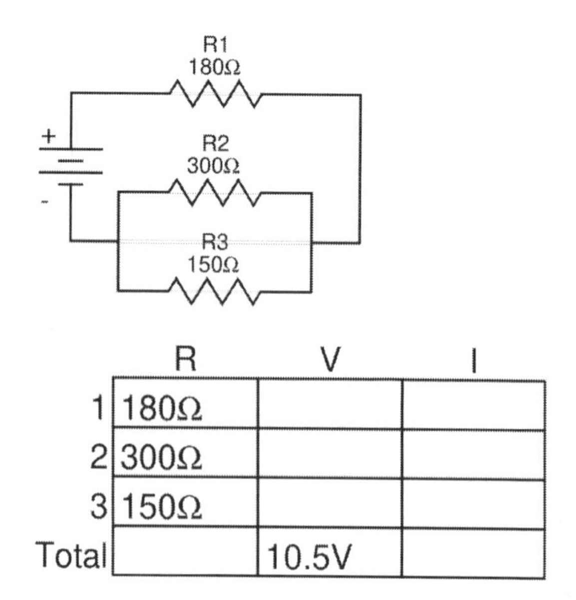

	R	V	I
1	180Ω		
2	300Ω		
3	150Ω		
Total		10.5V	

Exercise #4

Find the voltage and amperage for each resistor.

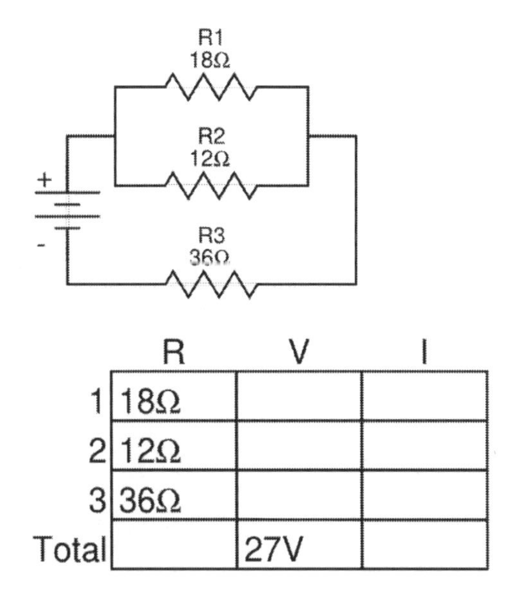

	R	V	I
1	18Ω		
2	12Ω		
3	36Ω		
Total		27V	

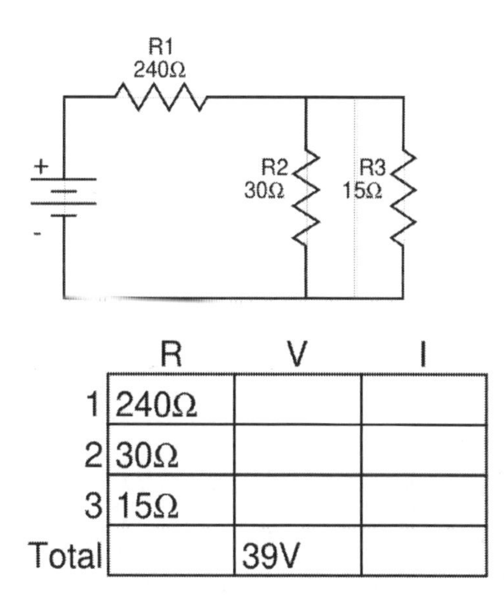

	R	V	I
1	240Ω		
2	30Ω		
3	15Ω		
Total		39V	

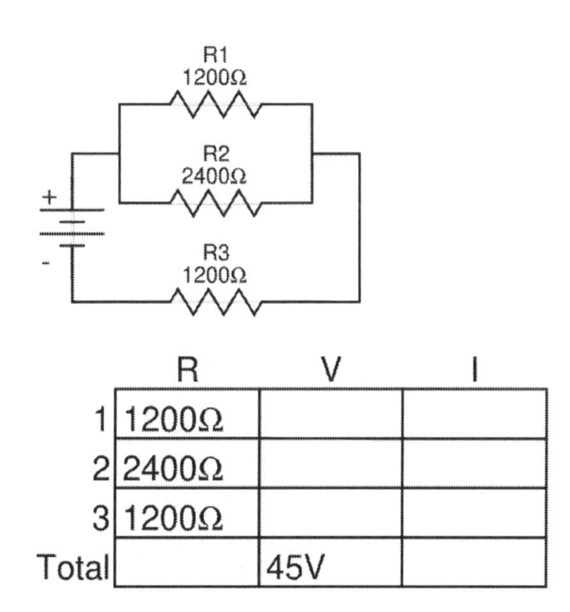

	R	V	I
1	1200Ω		
2	2400Ω		
3	1200Ω		
Total		45V	

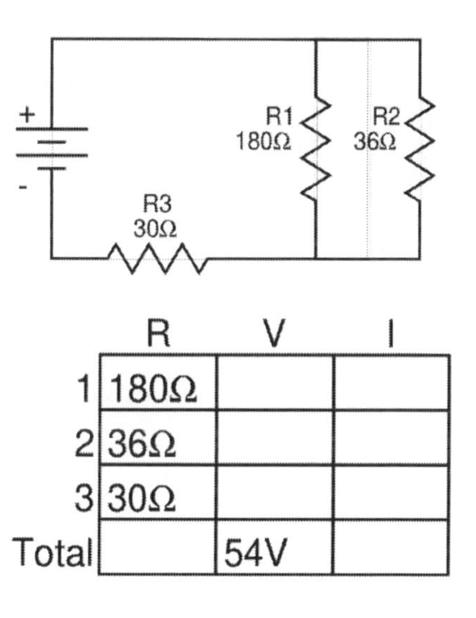

	R	V	I
1	180Ω		
2	36Ω		
3	30Ω		
Total		54V	

Exercise #5

Find the voltage and amperage for each resistor.

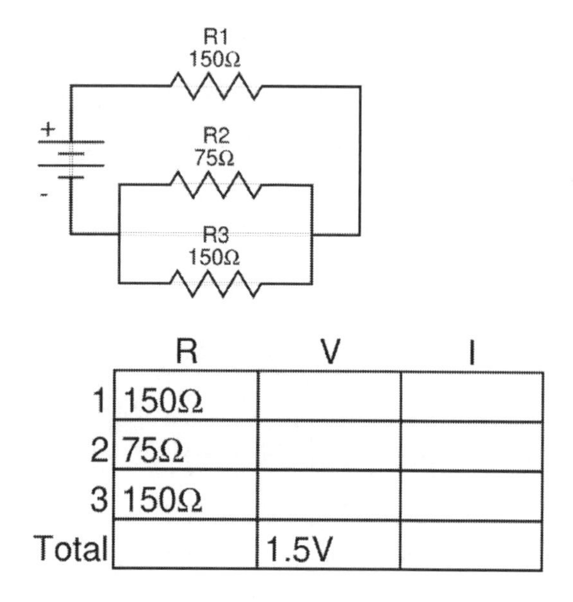

	R	V	I
1	150Ω		
2	75Ω		
3	150Ω		
Total		1.5V	

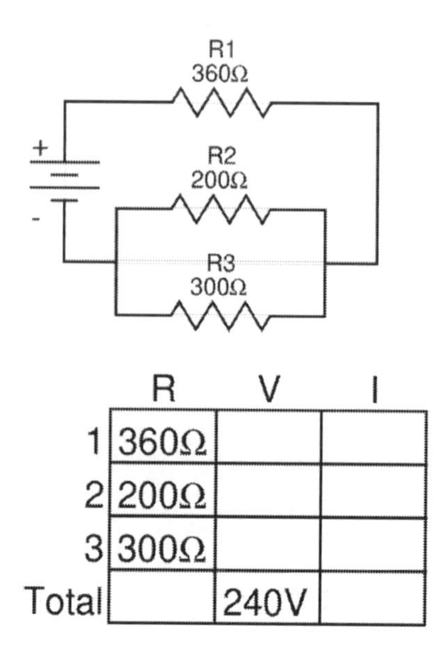

	R	V	I
1	360Ω		
2	200Ω		
3	300Ω		
Total		240V	

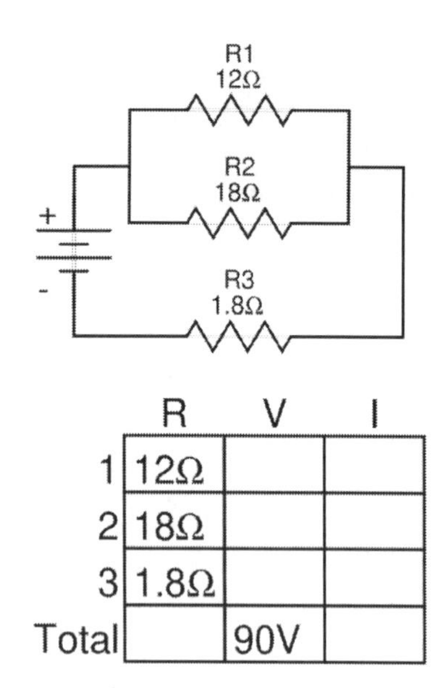

	R	V	I
1	12Ω		
2	18Ω		
3	1.8Ω		
Total		90V	

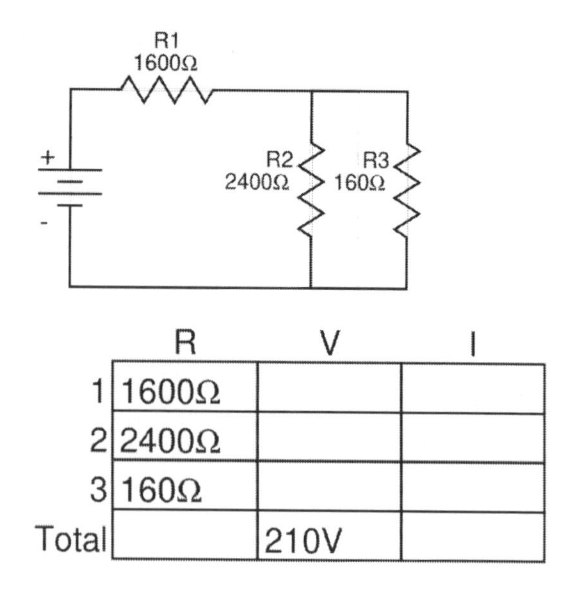

	R	V	I
1	1600Ω		
2	2400Ω		
3	160Ω		
Total		210V	

Exercise #6

Find the voltage and amperage for each resistor.

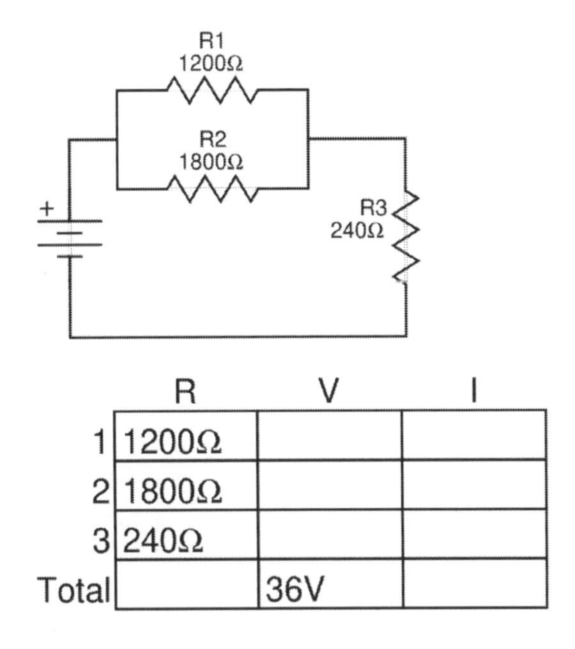

	R	V	I
1	1200Ω		
2	1800Ω		
3	240Ω		
Total		36V	

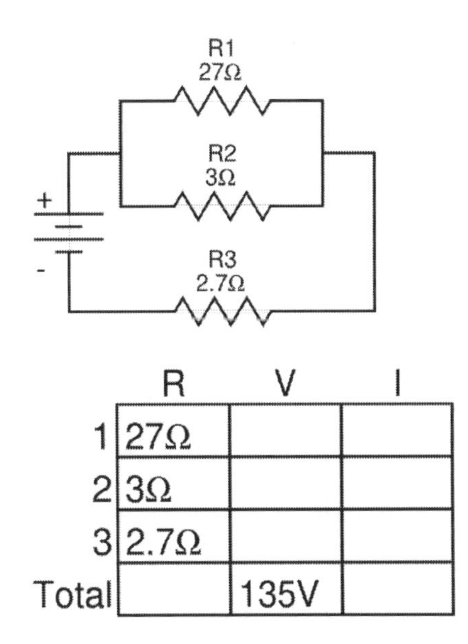

	R	V	I
1	27Ω		
2	3Ω		
3	2.7Ω		
Total		135V	

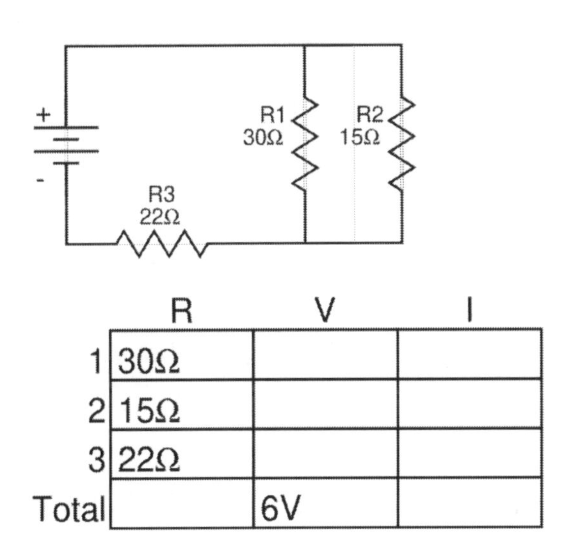

	R	V	I
1	30Ω		
2	15Ω		
3	22Ω		
Total		6V	

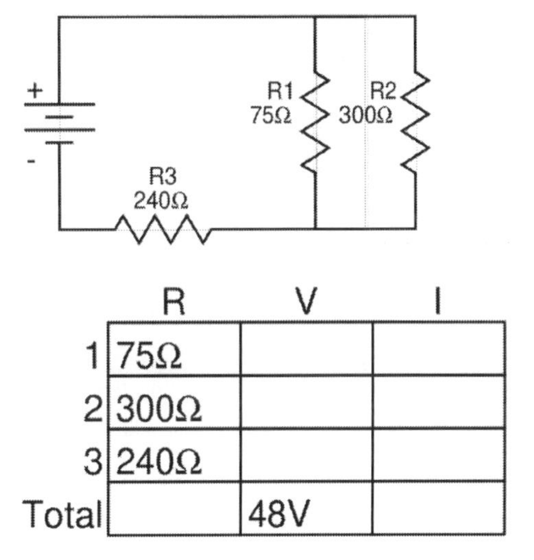

	R	V	I
1	75Ω		
2	300Ω		
3	240Ω		
Total		48V	

Exercise #7

Find the voltage and amperage for each resistor.

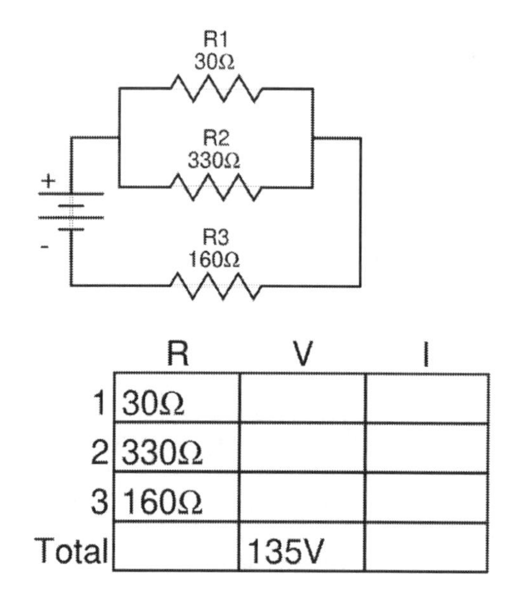

	R	V	I
1	30Ω		
2	330Ω		
3	160Ω		
Total		135V	

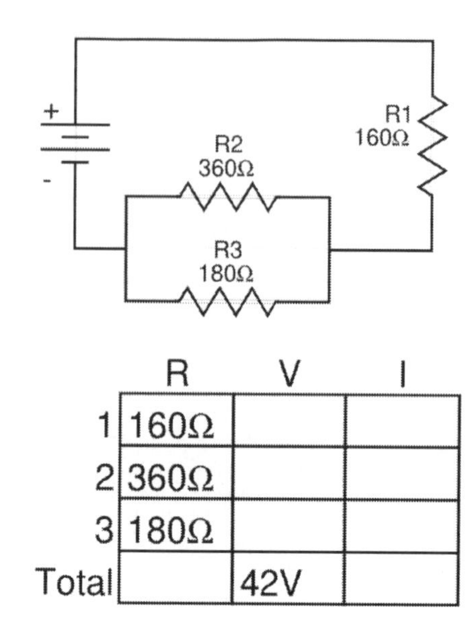

	R	V	I
1	160Ω		
2	360Ω		
3	180Ω		
Total		42V	

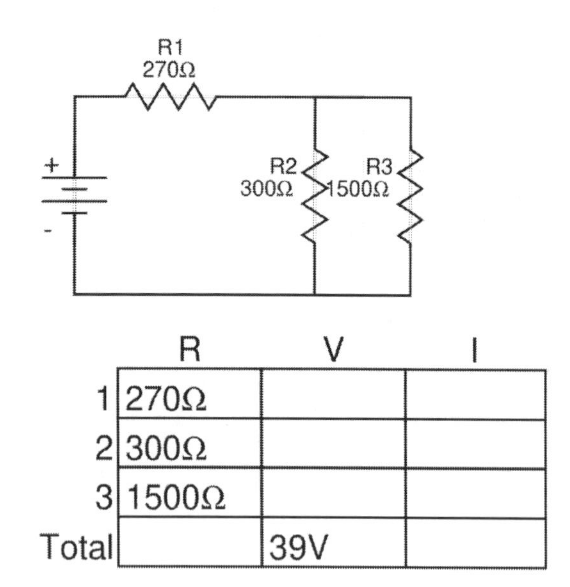

	R	V	I
1	270Ω		
2	300Ω		
3	1500Ω		
Total		39V	

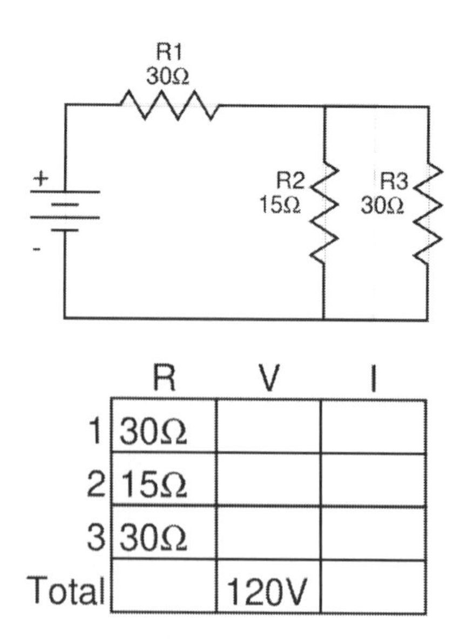

	R	V	I
1	30Ω		
2	15Ω		
3	30Ω		
Total		120V	

Exercise #8

Find the voltage and amperage for each resistor.

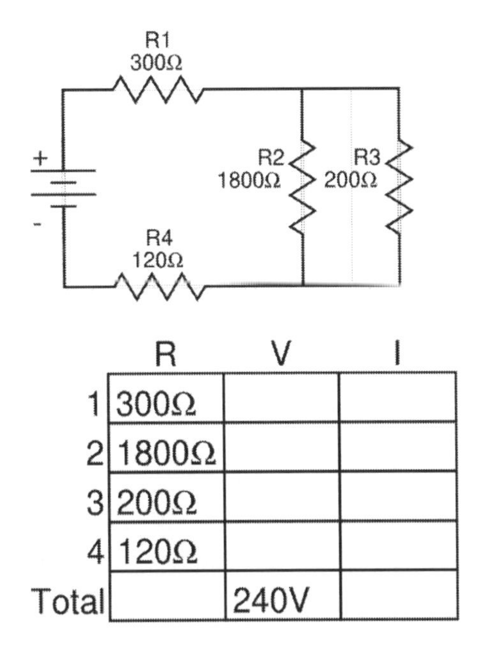

	R	V	I
1	300Ω		
2	1800Ω		
3	200Ω		
4	120Ω		
Total		240V	

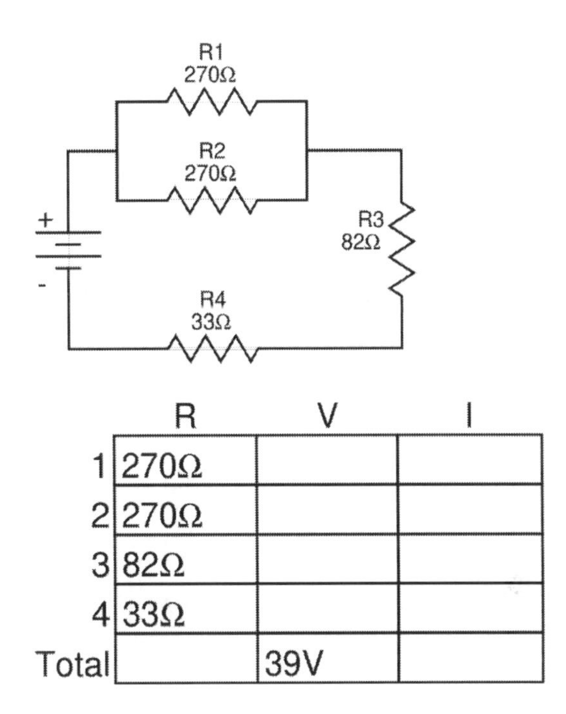

	R	V	I
1	270Ω		
2	270Ω		
3	82Ω		
4	33Ω		
Total		39V	

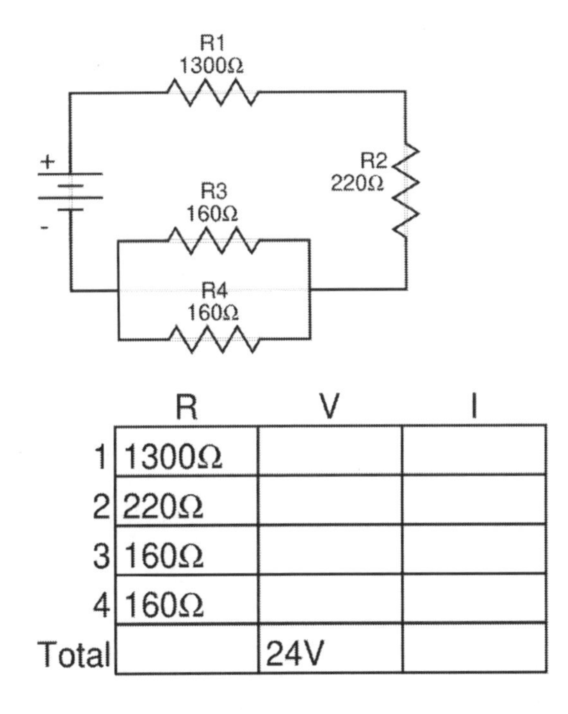

	R	V	I
1	1300Ω		
2	220Ω		
3	160Ω		
4	160Ω		
Total		24V	

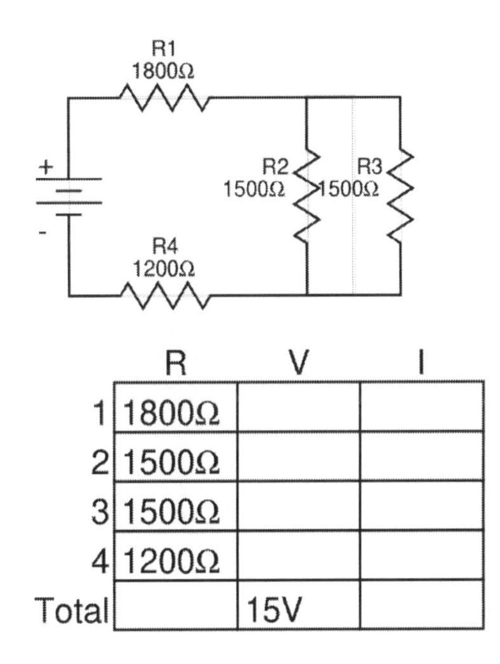

	R	V	I
1	1800Ω		
2	1500Ω		
3	1500Ω		
4	1200Ω		
Total		15V	

Exercise #9

Find the voltage and amperage for each resistor.

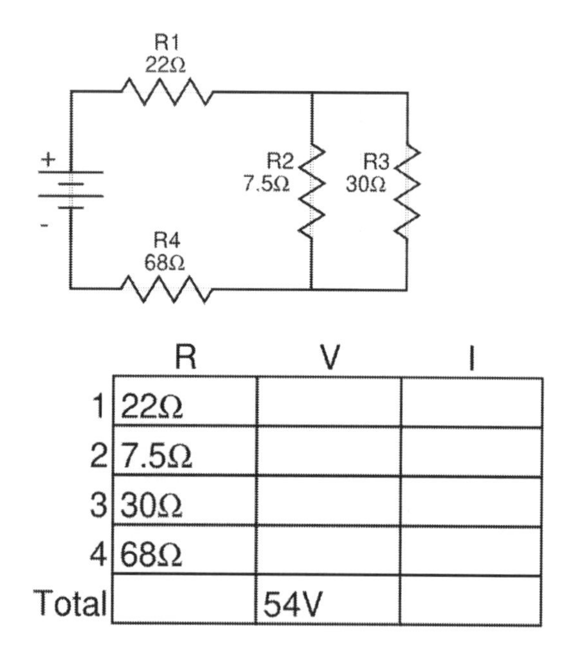

	R	V	I
1	22Ω		
2	7.5Ω		
3	30Ω		
4	68Ω		
Total		54V	

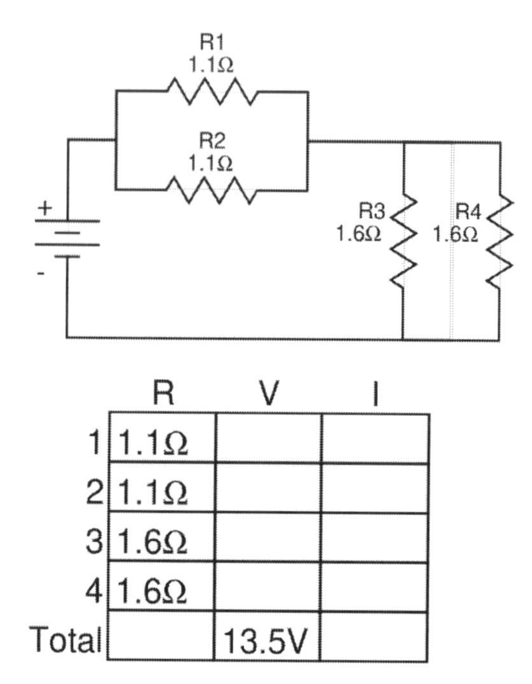

	R	V	I
1	1.1Ω		
2	1.1Ω		
3	1.6Ω		
4	1.6Ω		
Total		13.5V	

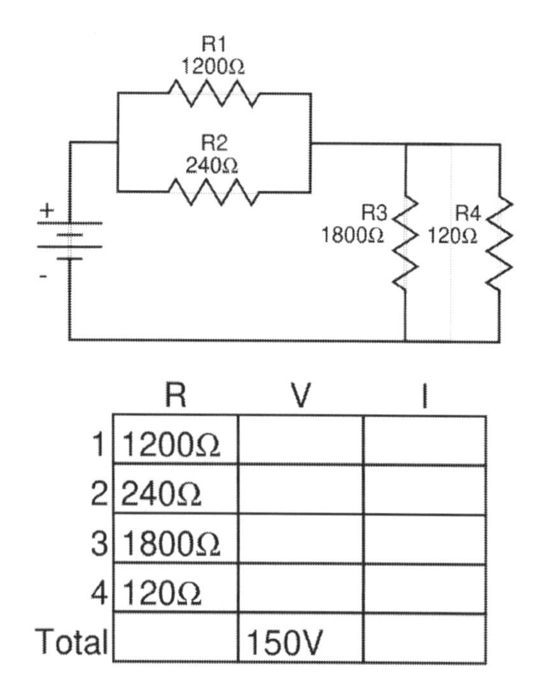

	R	V	I
1	1200Ω		
2	240Ω		
3	1800Ω		
4	120Ω		
Total		150V	

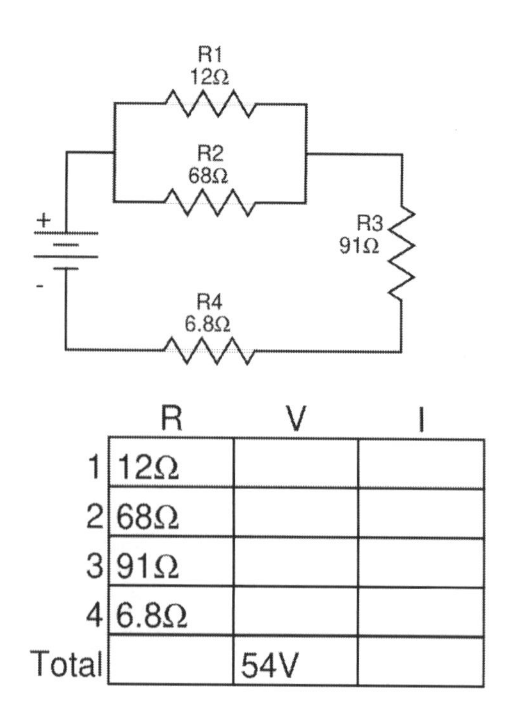

	R	V	I
1	12Ω		
2	68Ω		
3	91Ω		
4	6.8Ω		
Total		54V	

Exercise #10

Find the voltage and amperage for each resistor.

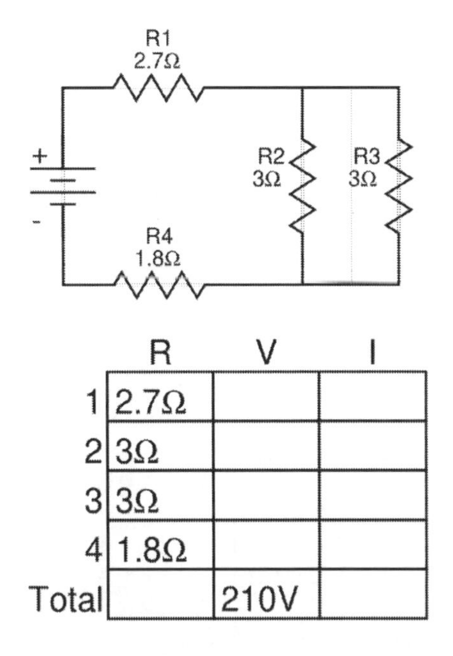

	R	V	I
1	2.7Ω		
2	3Ω		
3	3Ω		
4	1.8Ω		
Total		210V	

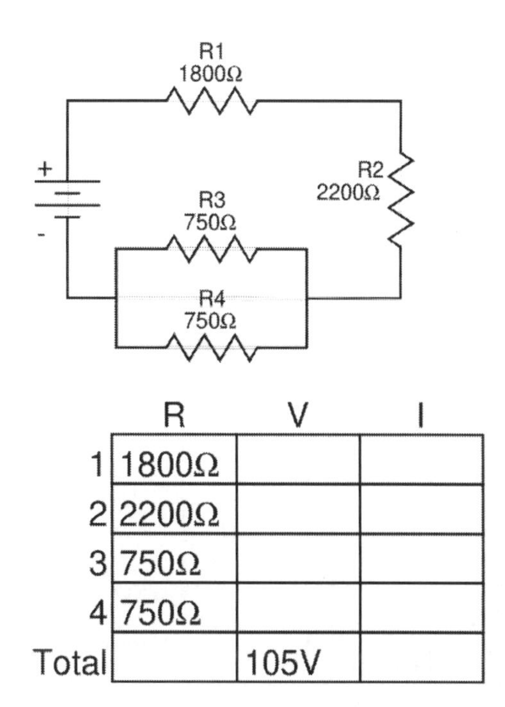

	R	V	I
1	1800Ω		
2	2200Ω		
3	750Ω		
4	750Ω		
Total		105V	

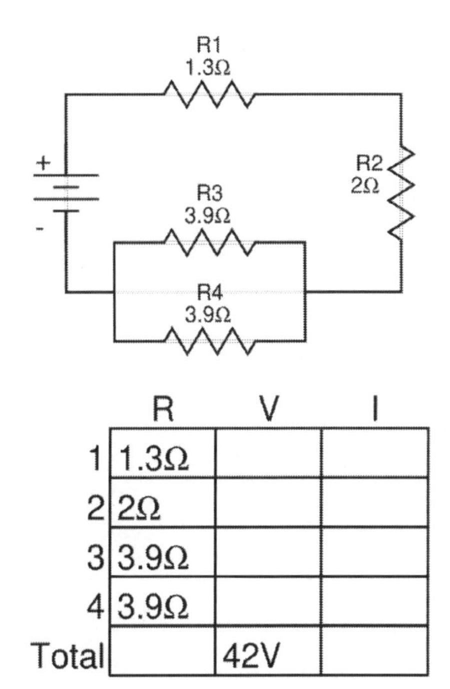

	R	V	I
1	1.3Ω		
2	2Ω		
3	3.9Ω		
4	3.9Ω		
Total		42V	

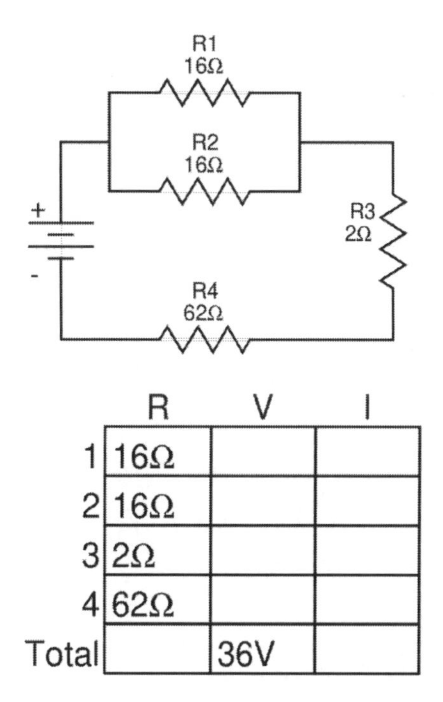

	R	V	I
1	16Ω		
2	16Ω		
3	2Ω		
4	62Ω		
Total		36V	

Exercise #11

Find the voltage and amperage for each resistor.

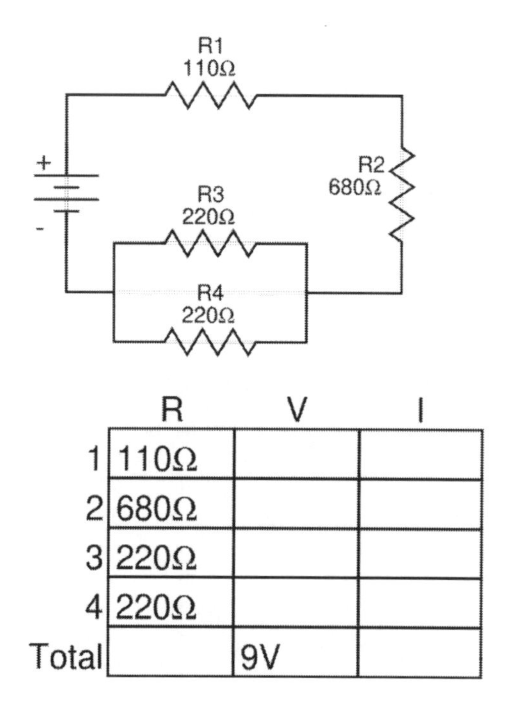

	R	V	I
1	110Ω		
2	680Ω		
3	220Ω		
4	220Ω		
Total		9V	

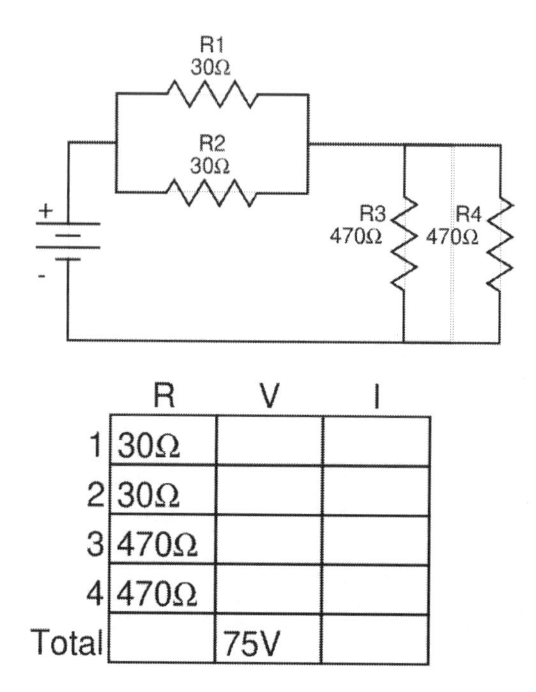

	R	V	I
1	30Ω		
2	30Ω		
3	470Ω		
4	470Ω		
Total		75V	

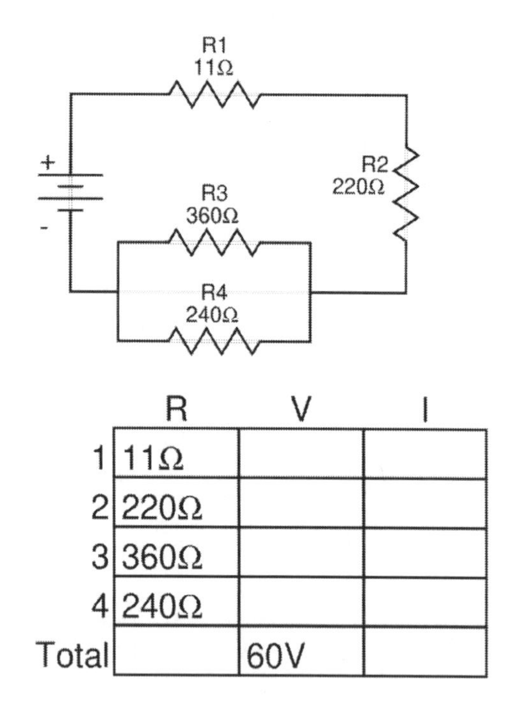

	R	V	I
1	11Ω		
2	220Ω		
3	360Ω		
4	240Ω		
Total		60V	

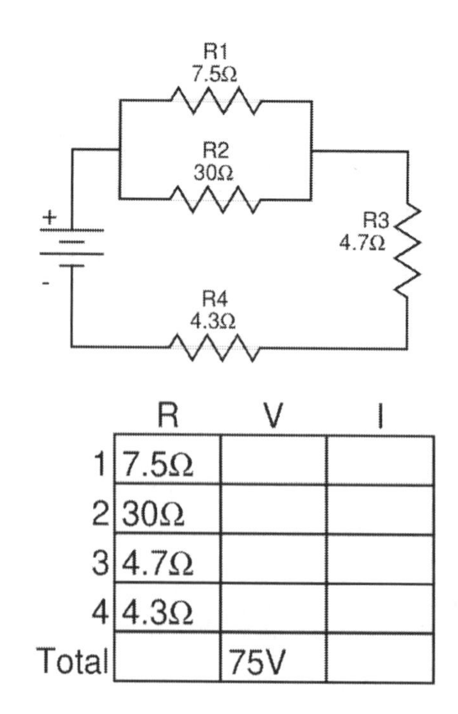

	R	V	I
1	7.5Ω		
2	30Ω		
3	4.7Ω		
4	4.3Ω		
Total		75V	

Exercise #12

Find the voltage and amperage for each resistor.

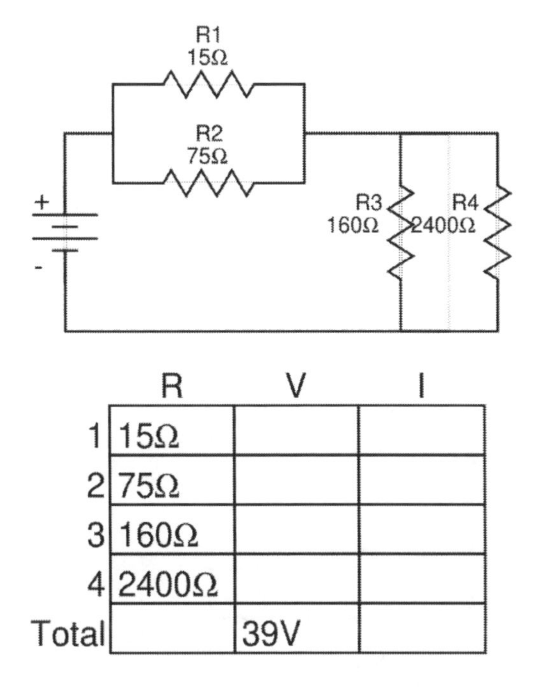

	R	V	I
1	15Ω		
2	75Ω		
3	160Ω		
4	2400Ω		
Total		39V	

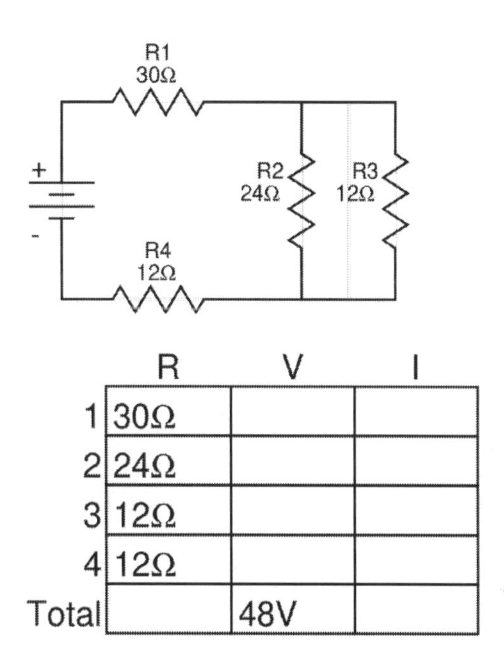

	R	V	I
1	30Ω		
2	24Ω		
3	12Ω		
4	12Ω		
Total		48V	

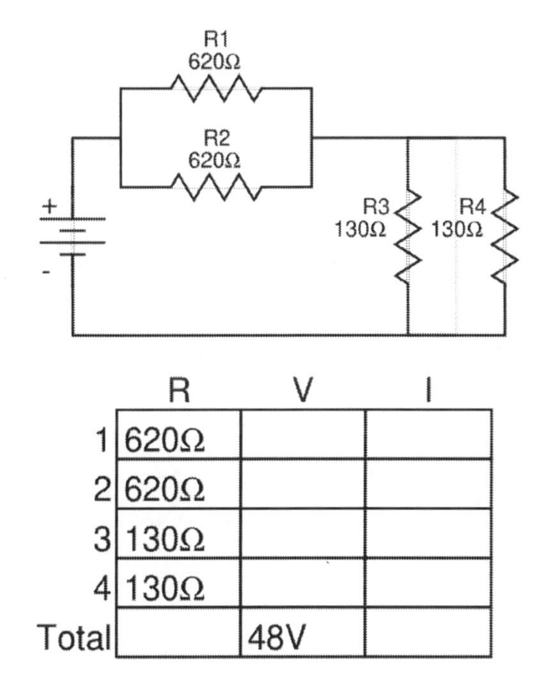

	R	V	I
1	620Ω		
2	620Ω		
3	130Ω		
4	130Ω		
Total		48V	

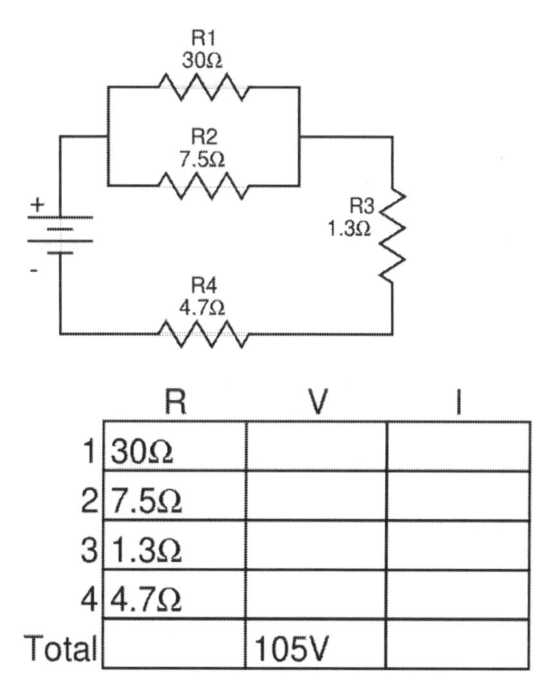

	R	V	I
1	30Ω		
2	7.5Ω		
3	1.3Ω		
4	4.7Ω		
Total		105V	

Exercise #13

Find the voltage and amperage for each resistor.

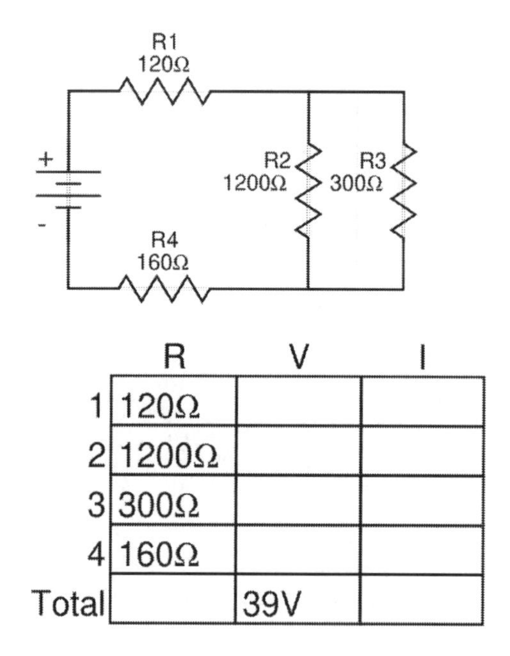

	R	V	I
1	120Ω		
2	1200Ω		
3	300Ω		
4	160Ω		
Total		39V	

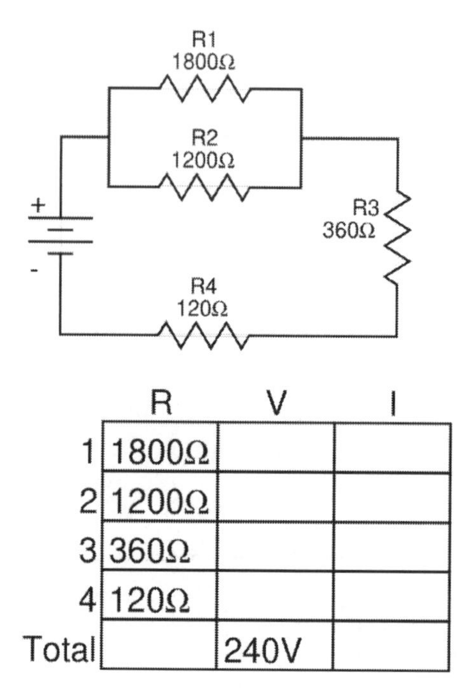

	R	V	I
1	1800Ω		
2	1200Ω		
3	360Ω		
4	120Ω		
Total		240V	

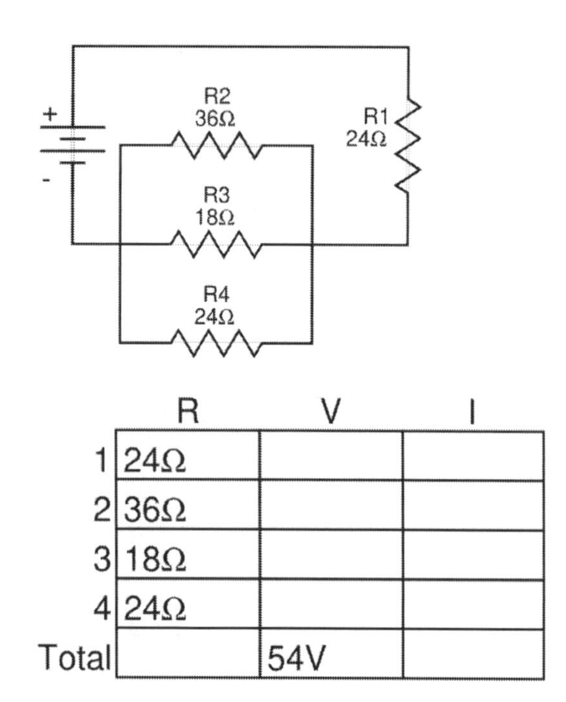

	R	V	I
1	24Ω		
2	36Ω		
3	18Ω		
4	24Ω		
Total		54V	

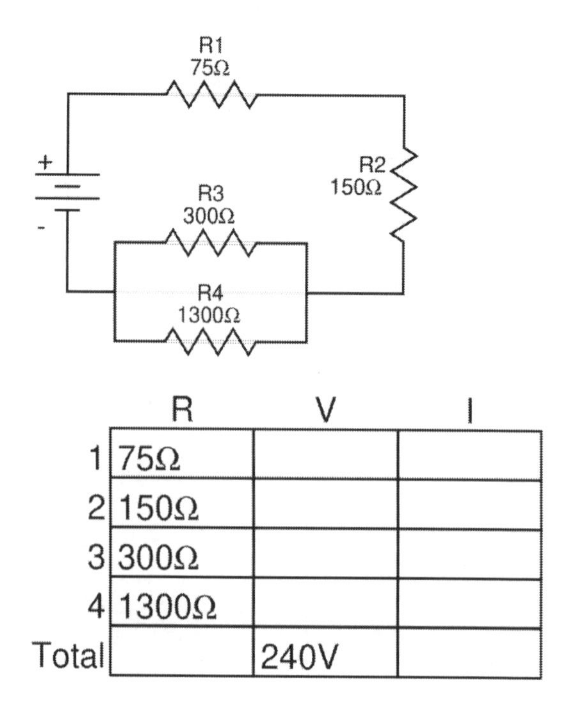

	R	V	I
1	75Ω		
2	150Ω		
3	300Ω		
4	1300Ω		
Total		240V	

Exercise #14

Find the voltage and amperage for each resistor.

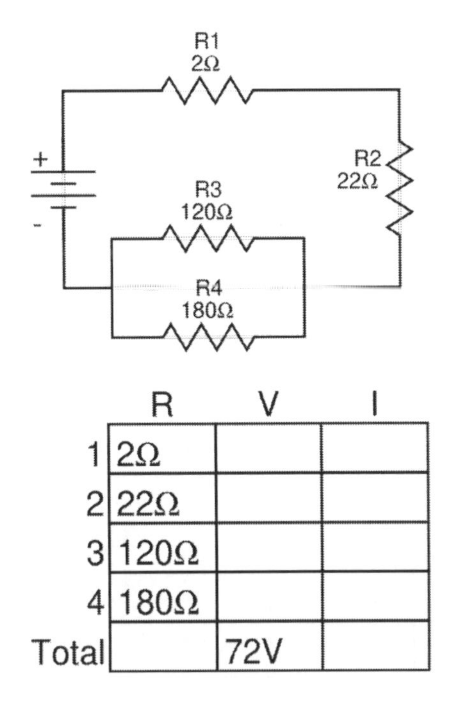

	R	V	I
1	2Ω		
2	22Ω		
3	120Ω		
4	180Ω		
Total		72V	

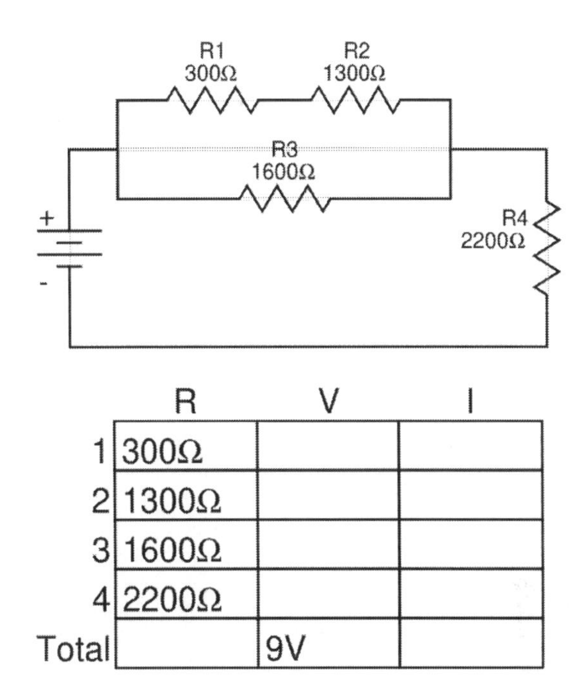

	R	V	I
1	300Ω		
2	1300Ω		
3	1600Ω		
4	2200Ω		
Total		9V	

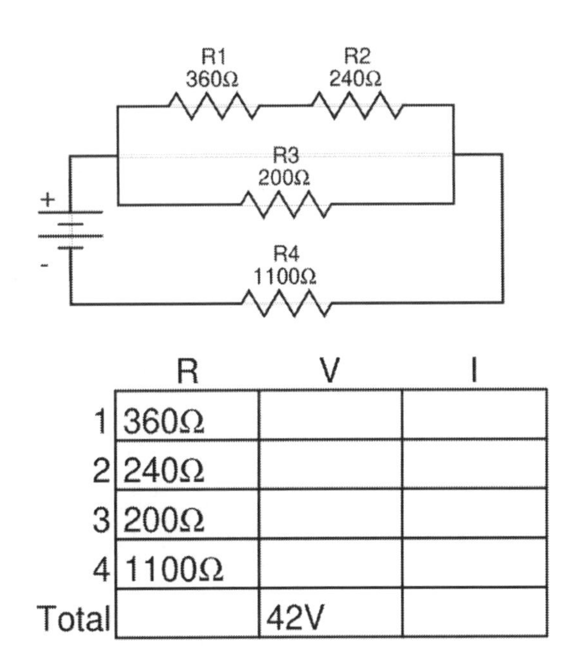

	R	V	I
1	360Ω		
2	240Ω		
3	200Ω		
4	1100Ω		
Total		42V	

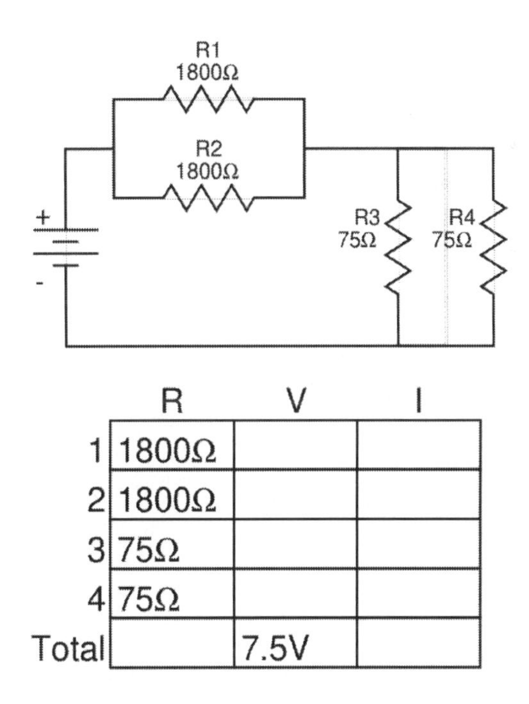

	R	V	I
1	1800Ω		
2	1800Ω		
3	75Ω		
4	75Ω		
Total		7.5V	

Exercise #15

Find the voltage and amperage for each resistor.

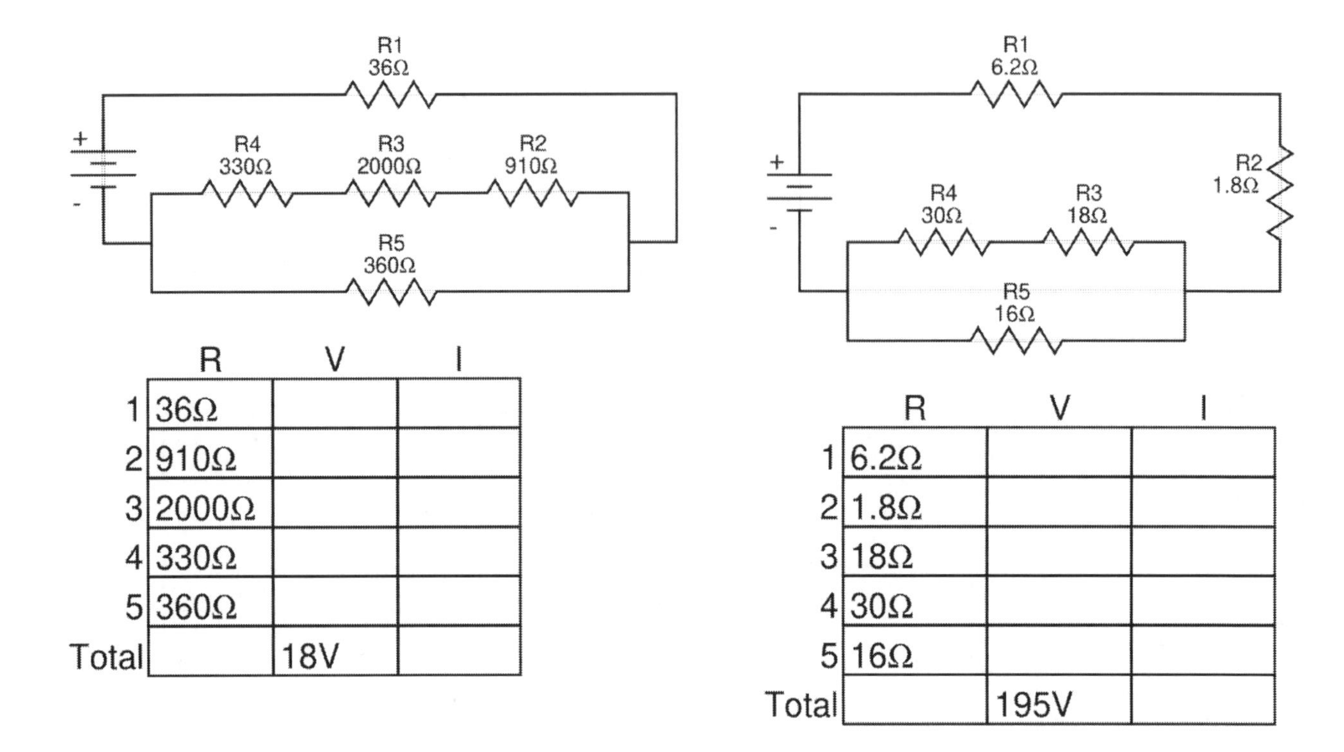

	R	V	I
1	36Ω		
2	910Ω		
3	2000Ω		
4	330Ω		
5	360Ω		
Total		18V	

	R	V	I
1	6.2Ω		
2	1.8Ω		
3	18Ω		
4	30Ω		
5	16Ω		
Total		195V	

Exercise #16

Find the voltage and amperage for each resistor.

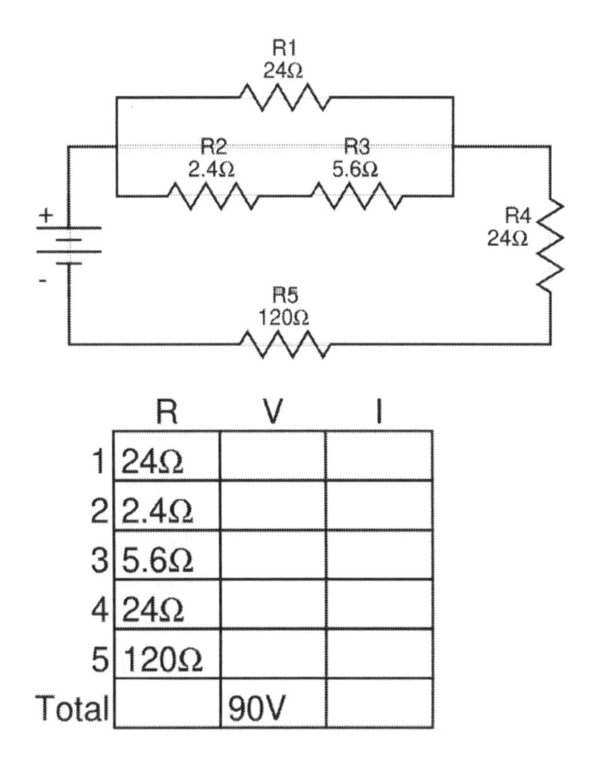

	R	V	I
1	24Ω		
2	2.4Ω		
3	5.6Ω		
4	24Ω		
5	120Ω		
Total		90V	

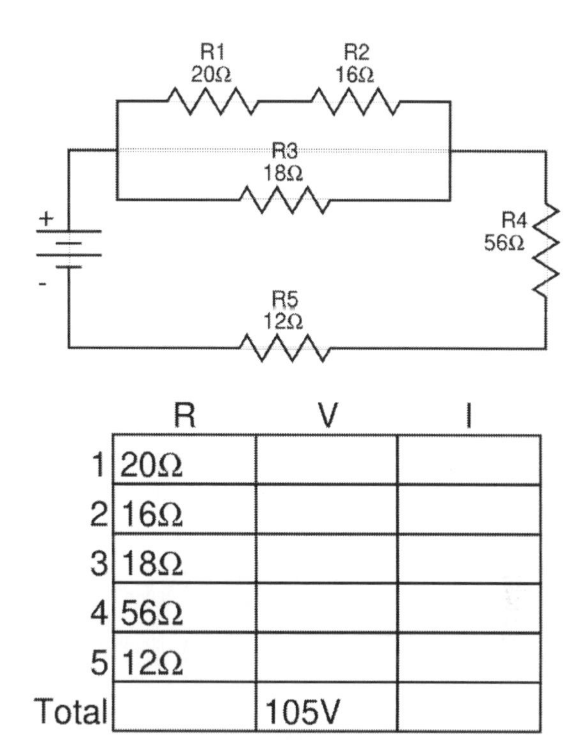

	R	V	I
1	20Ω		
2	16Ω		
3	18Ω		
4	56Ω		
5	12Ω		
Total		105V	

Exercise #17

Find the voltage and amperage for each resistor.

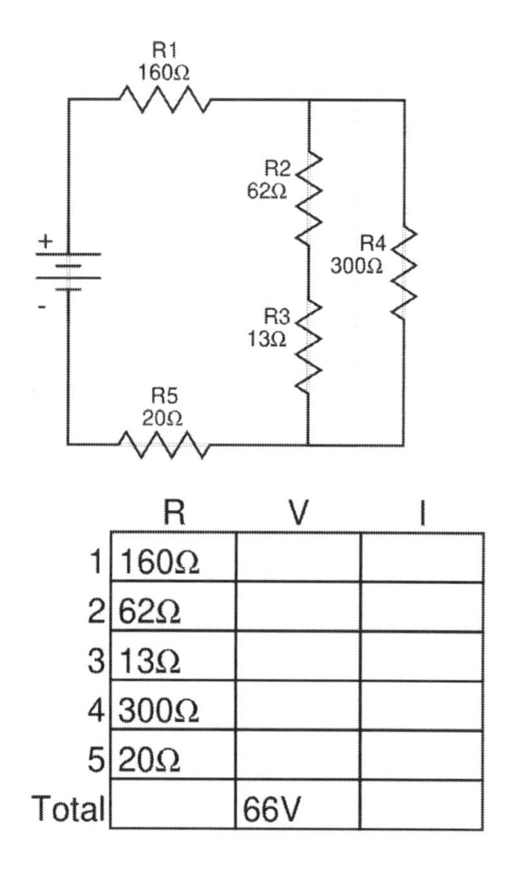

	R	V	I
1	160Ω		
2	62Ω		
3	13Ω		
4	300Ω		
5	20Ω		
Total		66V	

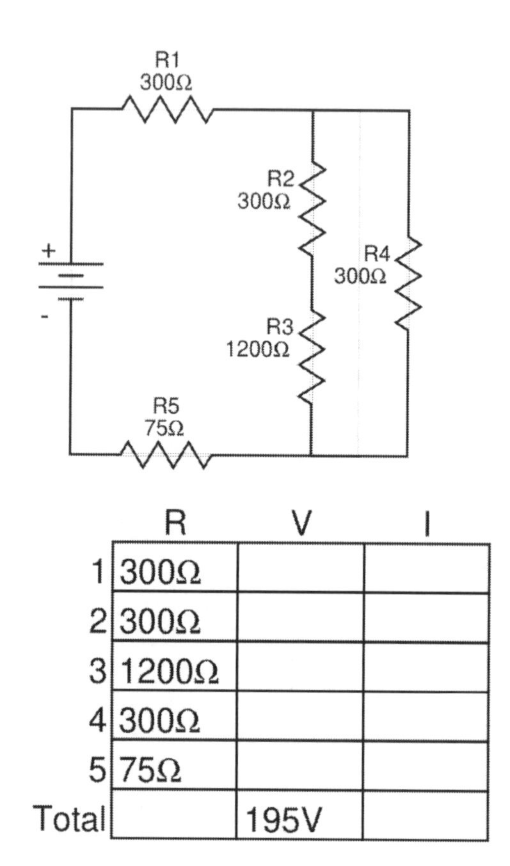

	R	V	I
1	300Ω		
2	300Ω		
3	1200Ω		
4	300Ω		
5	75Ω		
Total		195V	

Exercise #18

Find the voltage and amperage for each resistor.

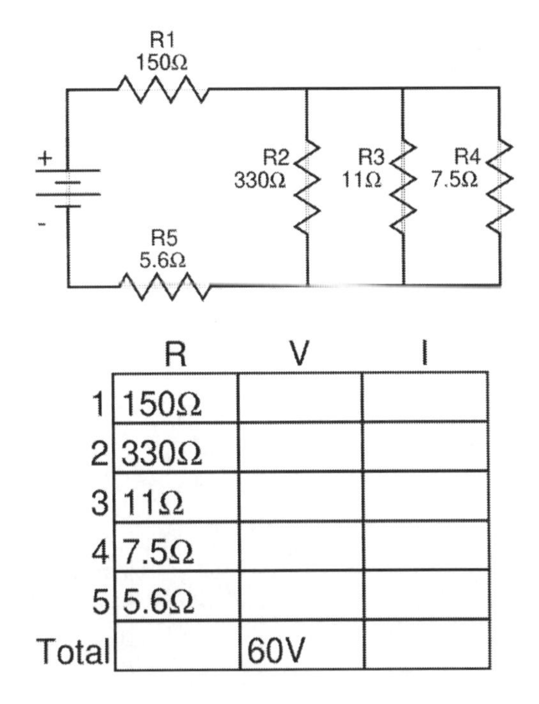

	R	V	I
1	150Ω		
2	330Ω		
3	11Ω		
4	7.5Ω		
5	5.6Ω		
Total		60V	

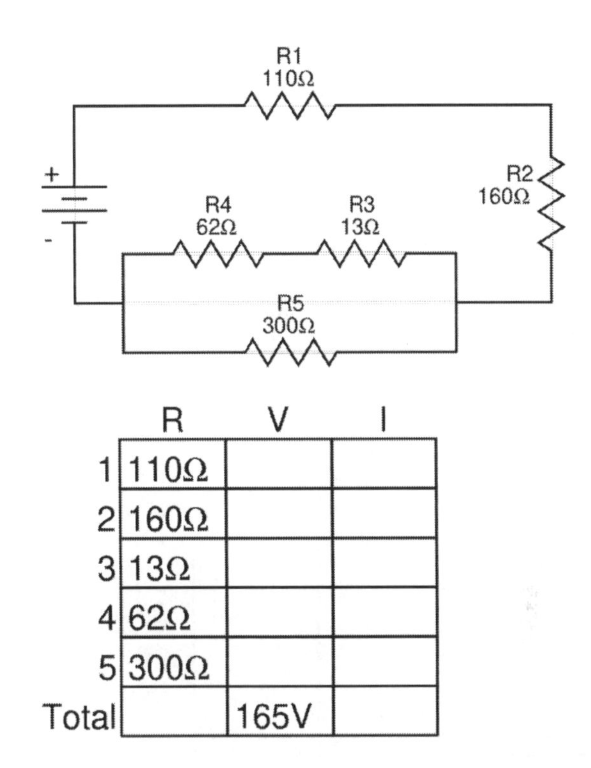

	R	V	I
1	110Ω		
2	160Ω		
3	13Ω		
4	62Ω		
5	300Ω		
Total		165V	

Exercise #19

Find the voltage and amperage for each resistor.

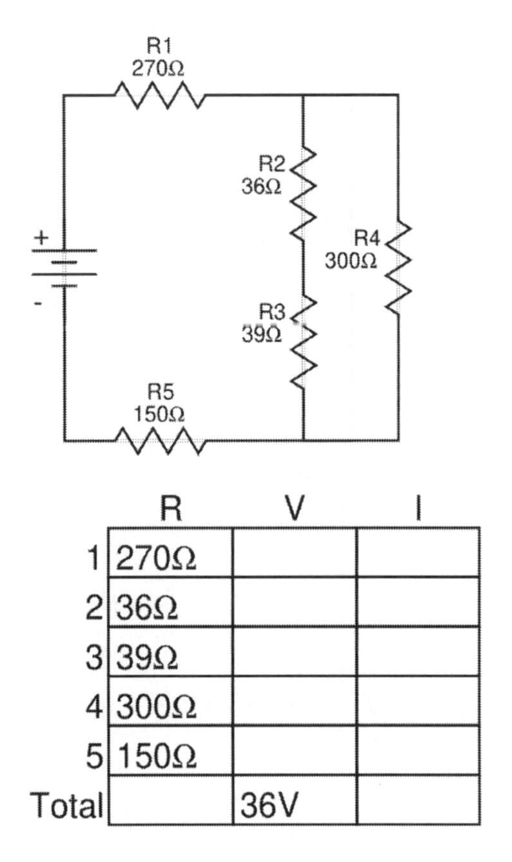

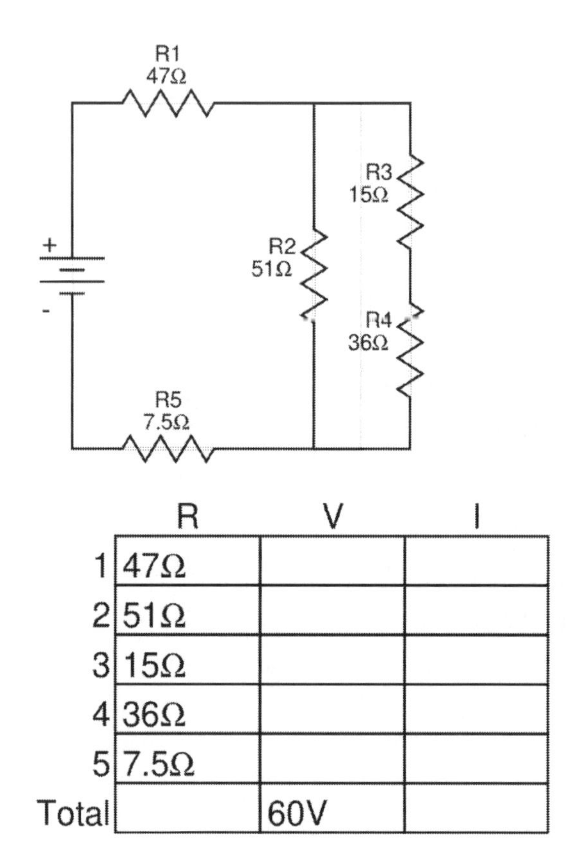

	R	V	I
1	270Ω		
2	36Ω		
3	39Ω		
4	300Ω		
5	150Ω		
Total		36V	

	R	V	I
1	47Ω		
2	51Ω		
3	15Ω		
4	36Ω		
5	7.5Ω		
Total		60V	

Exercise #20

Find the voltage and amperage for each resistor.

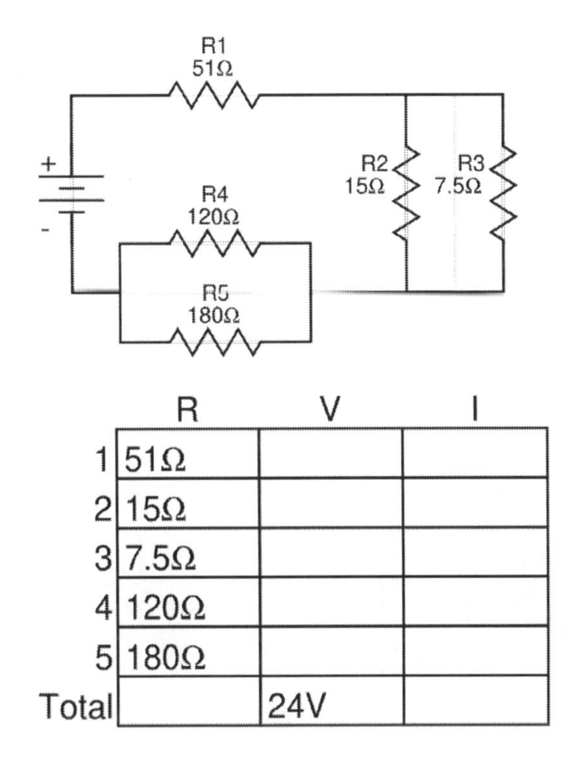

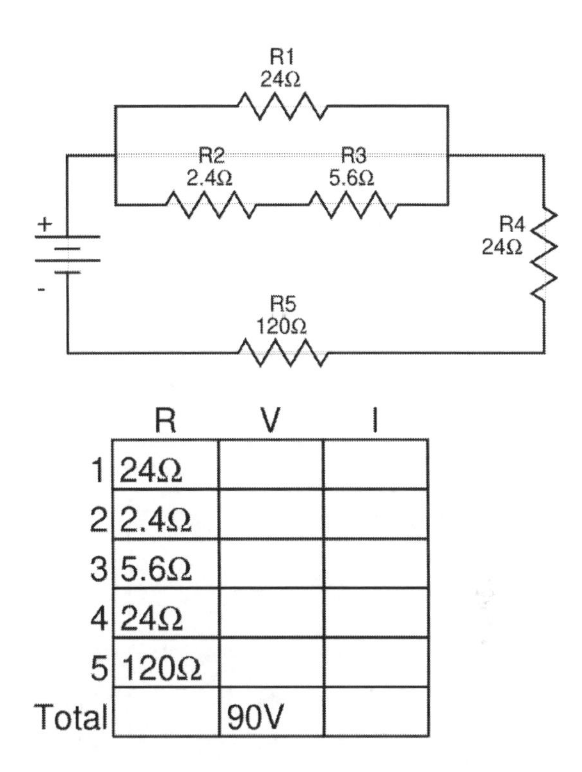

	R	V	I
1	51Ω		
2	15Ω		
3	7.5Ω		
4	120Ω		
5	180Ω		
Total		24V	

	R	V	I
1	24Ω		
2	2.4Ω		
3	5.6Ω		
4	24Ω		
5	120Ω		
Total		90V	

Exercise #21

Find the voltage and amperage for each resistor.

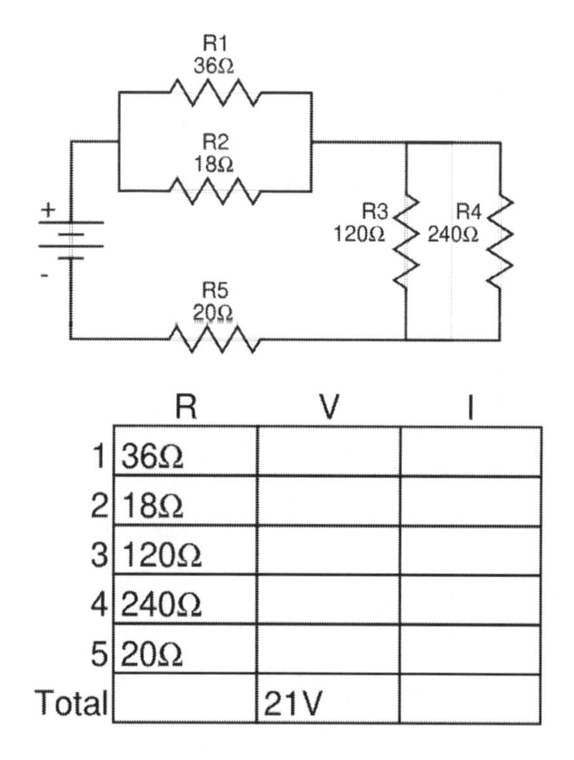

	R	V	I
1	36Ω		
2	18Ω		
3	120Ω		
4	240Ω		
5	20Ω		
Total		21V	

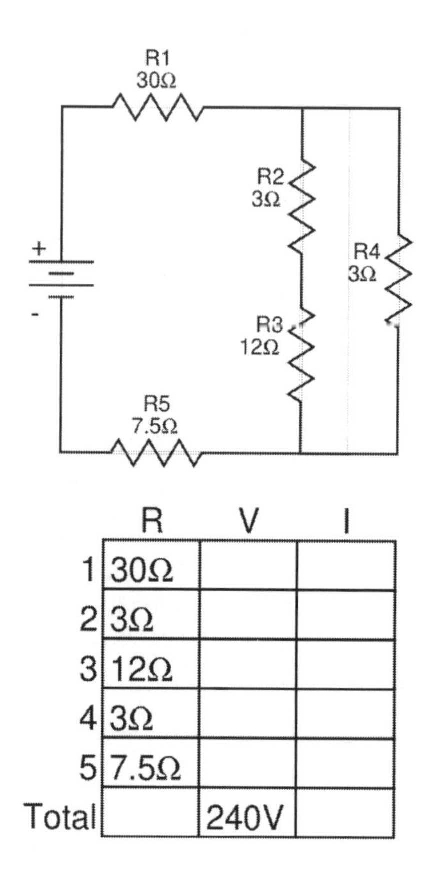

	R	V	I
1	30Ω		
2	3Ω		
3	12Ω		
4	3Ω		
5	7.5Ω		
Total		240V	

Exercise #22

Find the voltage and amperage for each resistor.

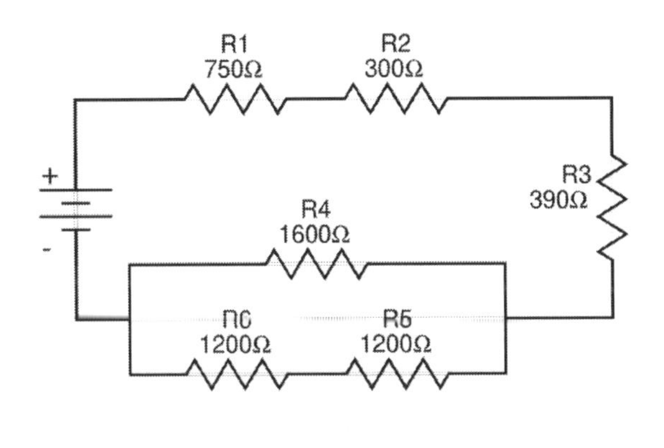

	R	V	I
1	750Ω		
2	300Ω		
3	390Ω		
4	1600Ω		
5	1200Ω		
6	1200Ω		
Total		180V	

Exercise #23

Find the voltage and amperage for each resistor.

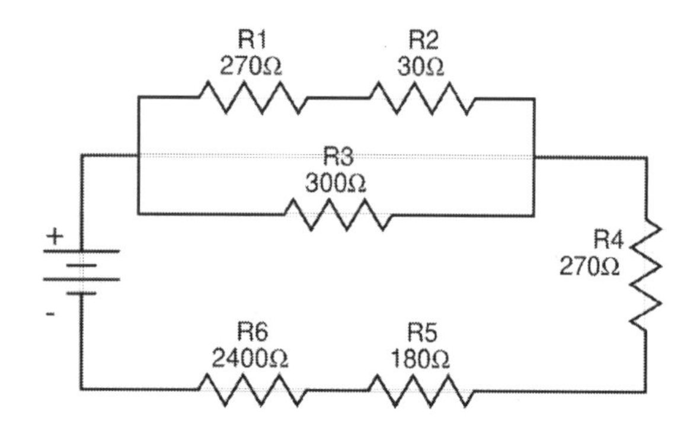

	R	V	I
1	270Ω		
2	30Ω		
3	300Ω		
4	270Ω		
5	180Ω		
6	2400Ω		
Total		3V	

Exercise #24

Find the voltage and amperage for each resistor.

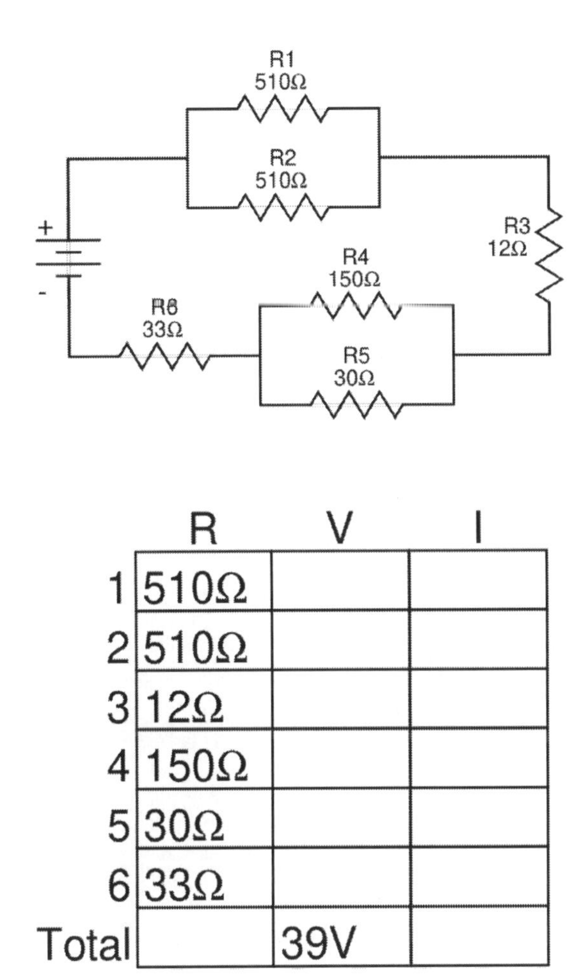

	R	V	I
1	510Ω		
2	510Ω		
3	12Ω		
4	150Ω		
5	30Ω		
6	33Ω		
Total		39V	

Exercise #25

Find the voltage and amperage for each resistor.

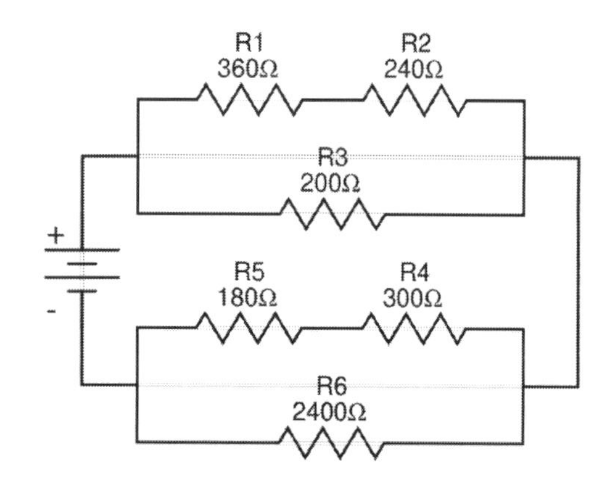

	R	V	I
1	360Ω		
2	240Ω		
3	200Ω		
4	300Ω		
5	180Ω		
6	2400Ω		
Total		165V	

Exercise #26

Find the voltage and amperage for each resistor.

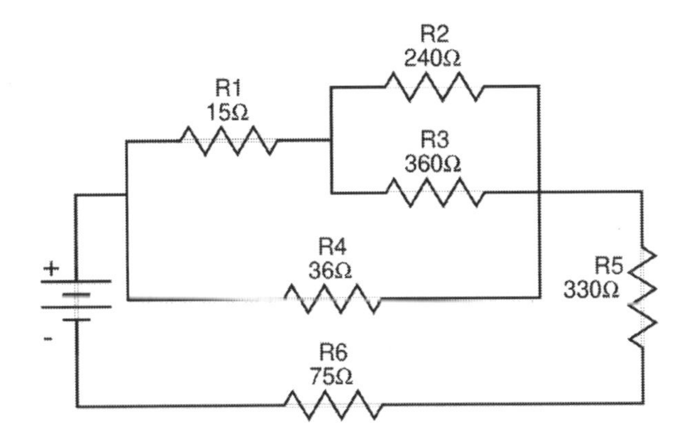

	R	V	I
1	15Ω		
2	240Ω		
3	360Ω		
4	36Ω		
5	330Ω		
6	75Ω		
Total		15V	

Exercise #27

Find the voltage and amperage for each resistor.

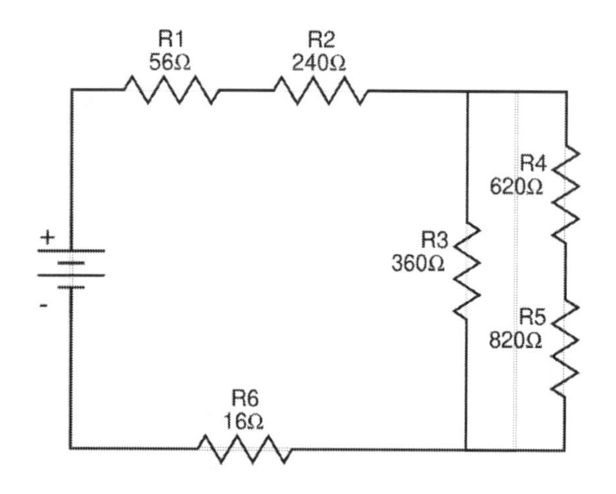

	R	V	I
1	56Ω		
2	240Ω		
3	360Ω		
4	620Ω		
5	820Ω		
6	16Ω		
Total		9V	

Exercise #28

Find the voltage and amperage for each resistor.

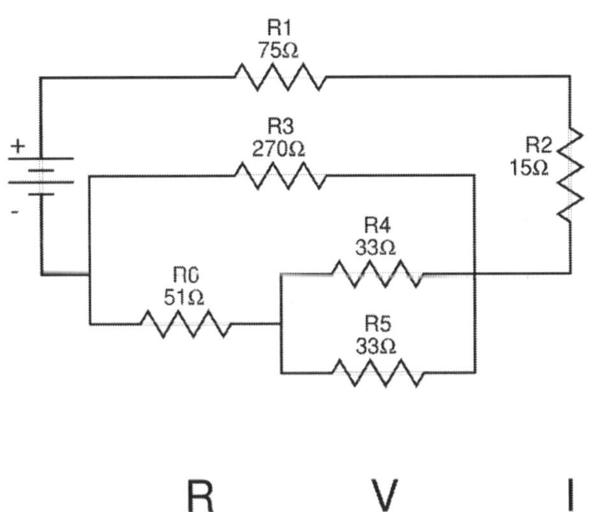

	R	V	I
1	75Ω		
2	15Ω		
3	270Ω		
4	33Ω		
5	33Ω		
6	51Ω		
Total		90V	

Exercise #29

Find the voltage and amperage for each resistor.

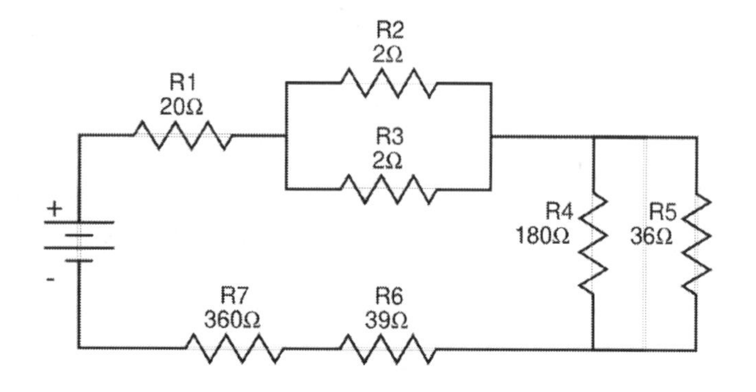

	R	V	I
1	20Ω		
2	2Ω		
3	2Ω		
4	180Ω		
5	36Ω		
6	39Ω		
7	360Ω		
Total		135V	

Exercise #30

Find the voltage and amperage for each resistor.

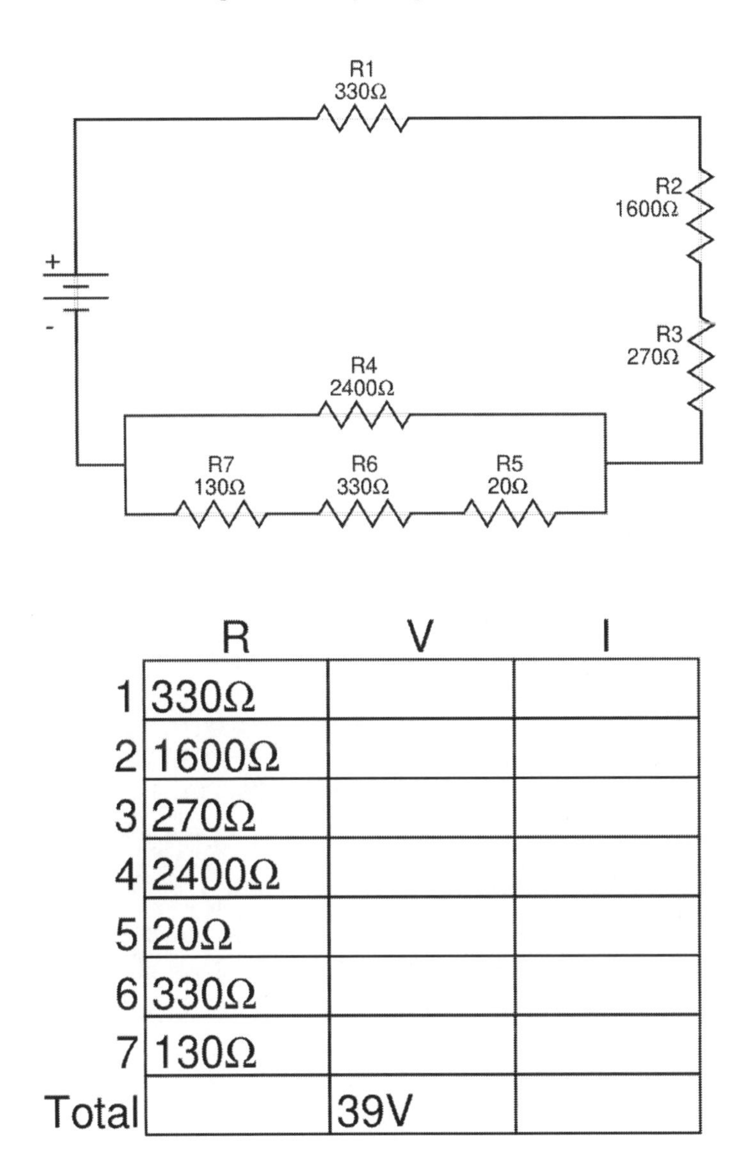

	R	V	I
1	330Ω		
2	1600Ω		
3	270Ω		
4	2400Ω		
5	20Ω		
6	330Ω		
7	130Ω		
Total		39V	

Exercise #31

Find the voltage and amperage for each resistor.

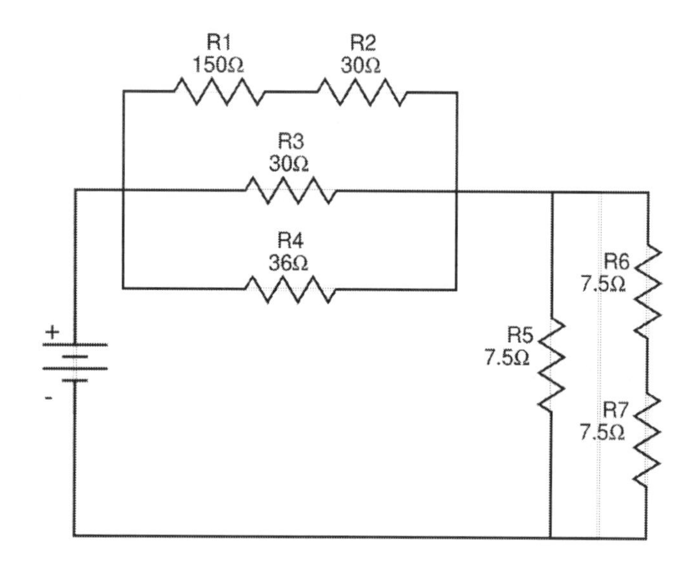

	R	V	I
1	150Ω		
2	30Ω		
3	30Ω		
4	36Ω		
5	7.5Ω		
6	7.5Ω		
7	7.5Ω		
Total		150V	

Exercise #32

Find the voltage and amperage for each resistor.

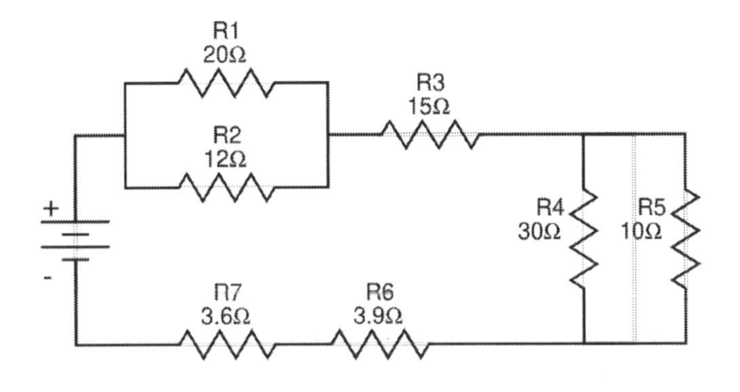

	R	V	I
1	20Ω		
2	12Ω		
3	15Ω		
4	30Ω		
5	10Ω		
6	3.9Ω		
7	3.6Ω		
Total		75V	

Exercise #33

Find the voltage and amperage for each resistor.

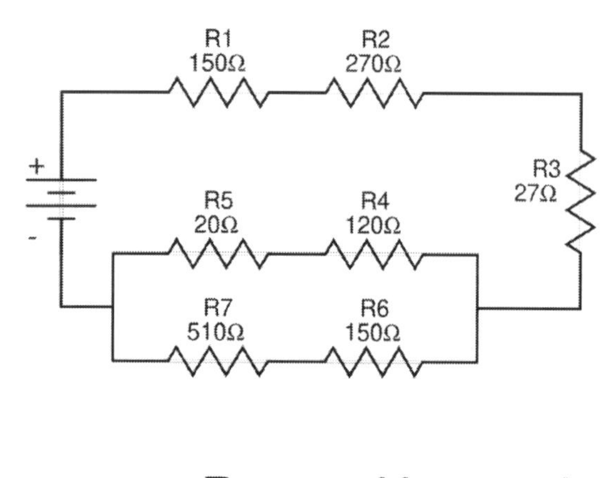

	R	V	I
1	150Ω		
2	270Ω		
3	27Ω		
4	120Ω		
5	20Ω		
6	150Ω		
7	510Ω		
Total		180V	

Exercise #34

Find the voltage and amperage for each resistor.

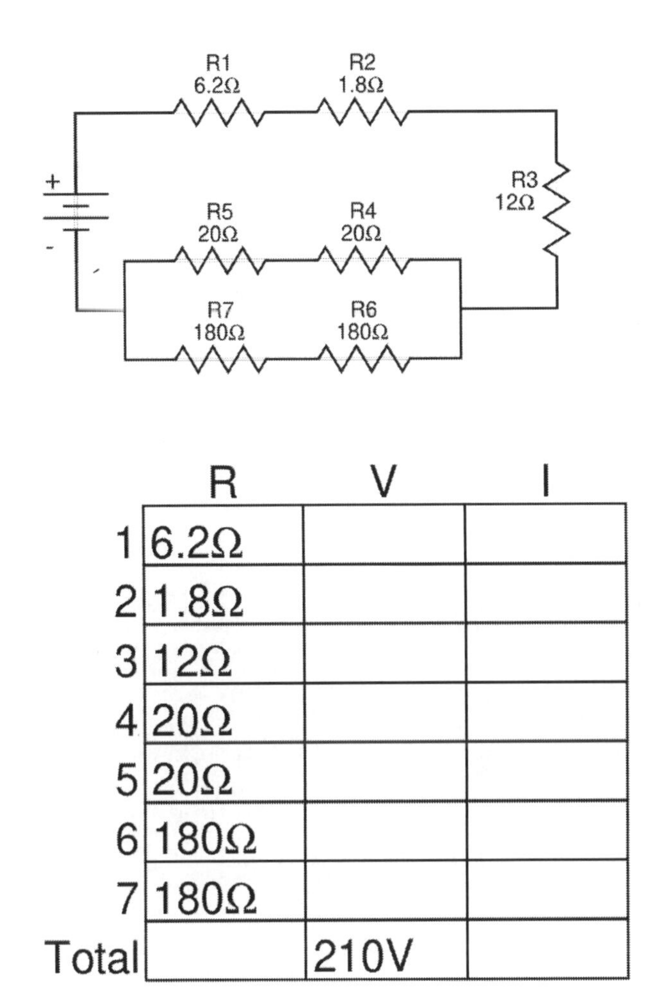

	R	V	I
1	6.2Ω		
2	1.8Ω		
3	12Ω		
4	20Ω		
5	20Ω		
6	180Ω		
7	180Ω		
Total		210V	

Exercise #35

Find the voltage and amperage for each resistor.

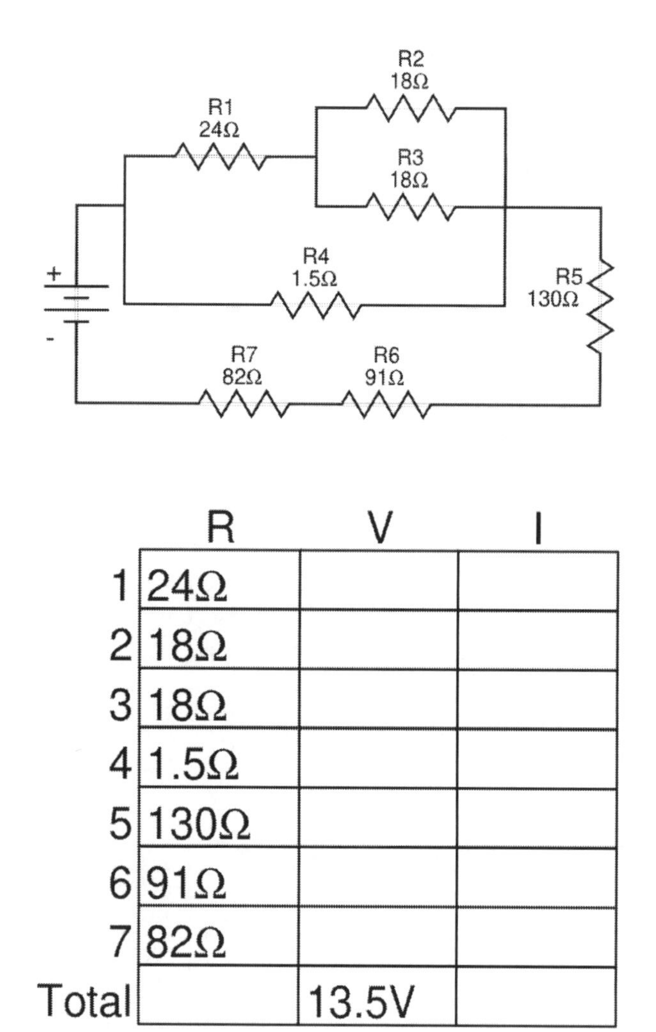

	R	V	I
1	24Ω		
2	18Ω		
3	18Ω		
4	1.5Ω		
5	130Ω		
6	91Ω		
7	82Ω		
Total		13.5V	

Exercise #36

Find the voltage and amperage for each resistor.

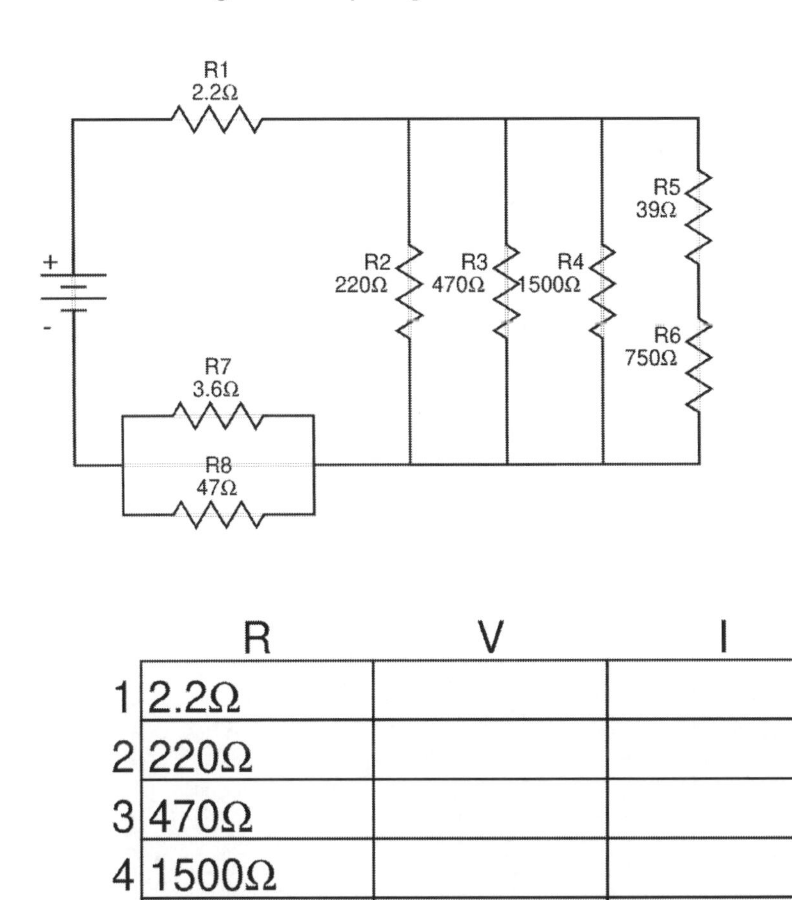

	R	V	I
1	2.2Ω		
2	220Ω		
3	470Ω		
4	1500Ω		
5	39Ω		
6	750Ω		
7	3.6Ω		
8	47Ω		
Total		6V	

Exercise #37

Find the voltage and amperage for each resistor.

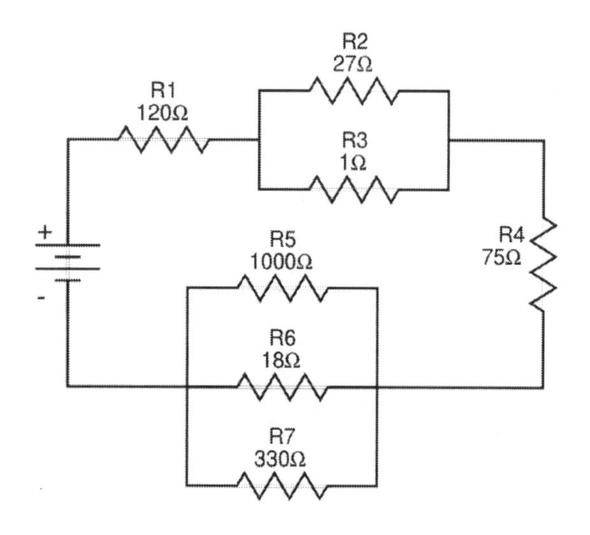

	R	V	I
1	120Ω		
2	27Ω		
3	1Ω		
4	75Ω		
5	1000Ω		
6	18Ω		
7	330Ω		
Total		9V	

Exercise #38

Find the voltage and amperage for each resistor.

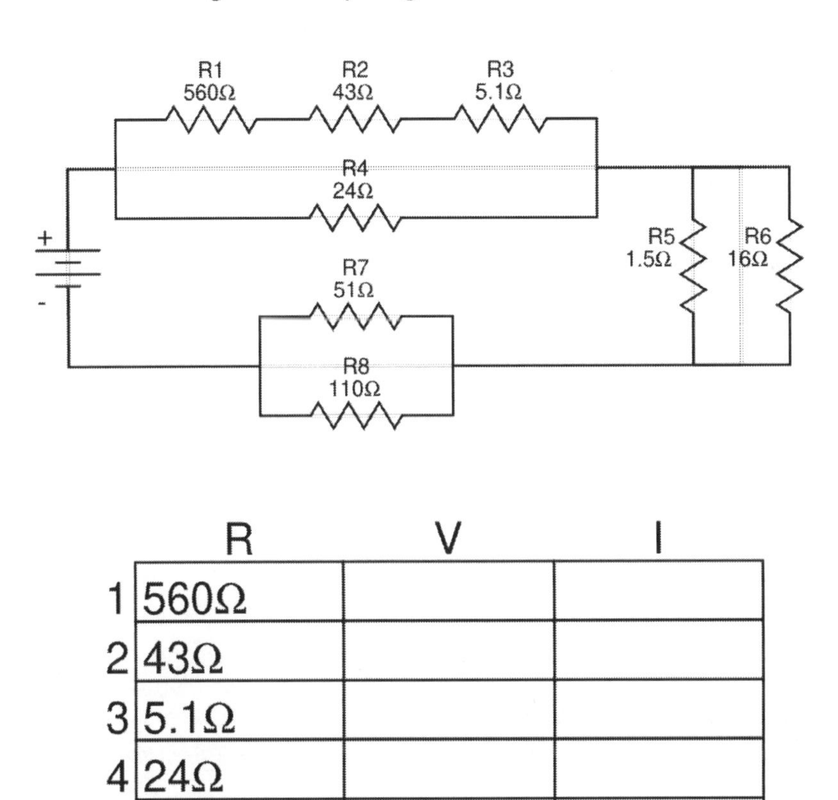

	R	V	I
1	560Ω		
2	43Ω		
3	5.1Ω		
4	24Ω		
5	1.5Ω		
6	16Ω		
7	51Ω		
8	110Ω		
Total		72V	

Exercise #39

Find the voltage and amperage for each resistor.

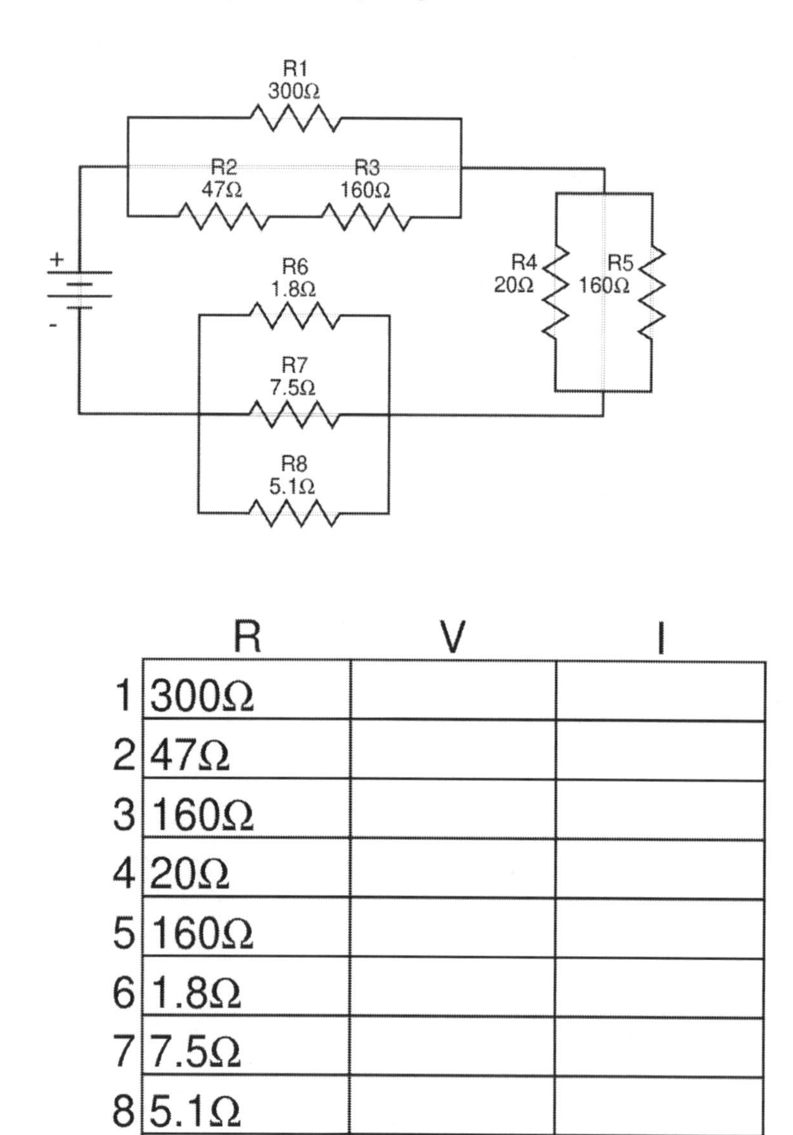

	R	V	I
1	300Ω		
2	47Ω		
3	160Ω		
4	20Ω		
5	160Ω		
6	1.8Ω		
7	7.5Ω		
8	5.1Ω		
Total		4.5V	

Exercise #40

Find the voltage and amperage for each resistor.

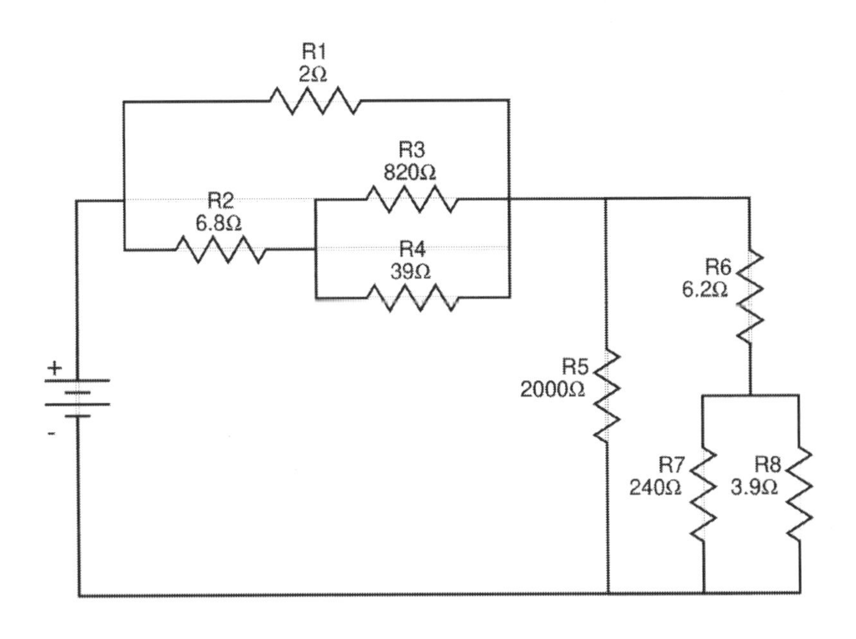

	R	V	I
1	2Ω		
2	6.8Ω		
3	820Ω		
4	39Ω		
5	2000Ω		
6	6.2Ω		
7	240Ω		
8	3.9Ω		
Total		33V	

Exercise #41

Find the voltage and amperage for each resistor.

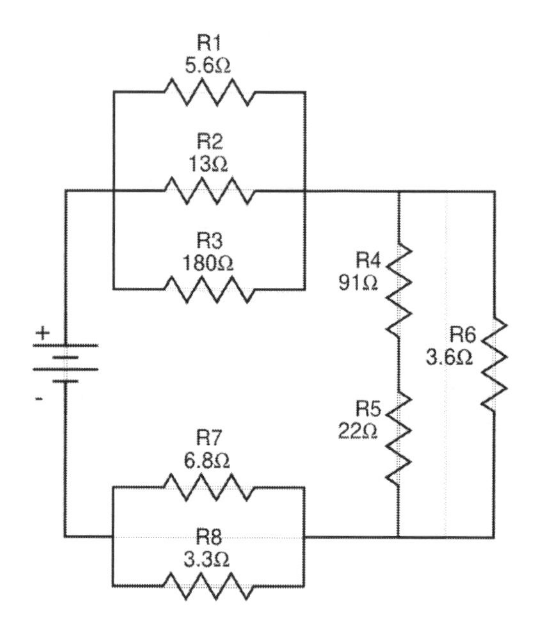

	R	V	I
1	5.6Ω		
2	13Ω		
3	180Ω		
4	91Ω		
5	22Ω		
6	3.6Ω		
7	6.8Ω		
8	3.3Ω		
Total		105V	

Exercise #42

Find the voltage and amperage for each resistor.

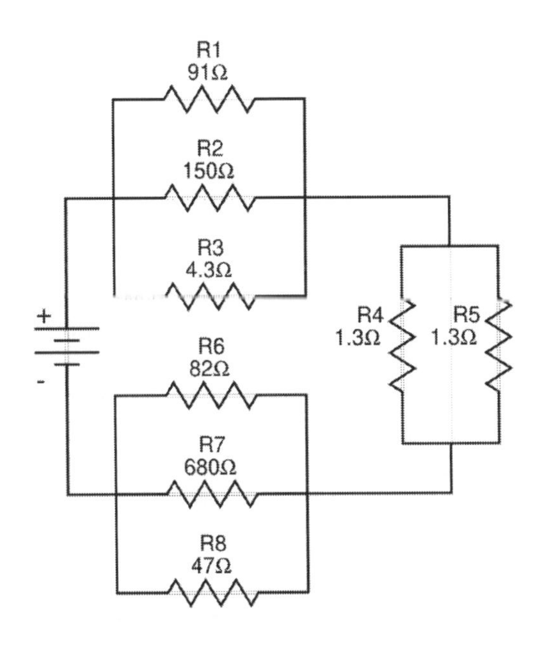

	R	V	I
1	91Ω		
2	150Ω		
3	4.3Ω		
4	1.3Ω		
5	1.3Ω		
6	82Ω		
7	680Ω		
8	47Ω		
Total		150V	

Exercise #43

Find the voltage and amperage for each resistor.

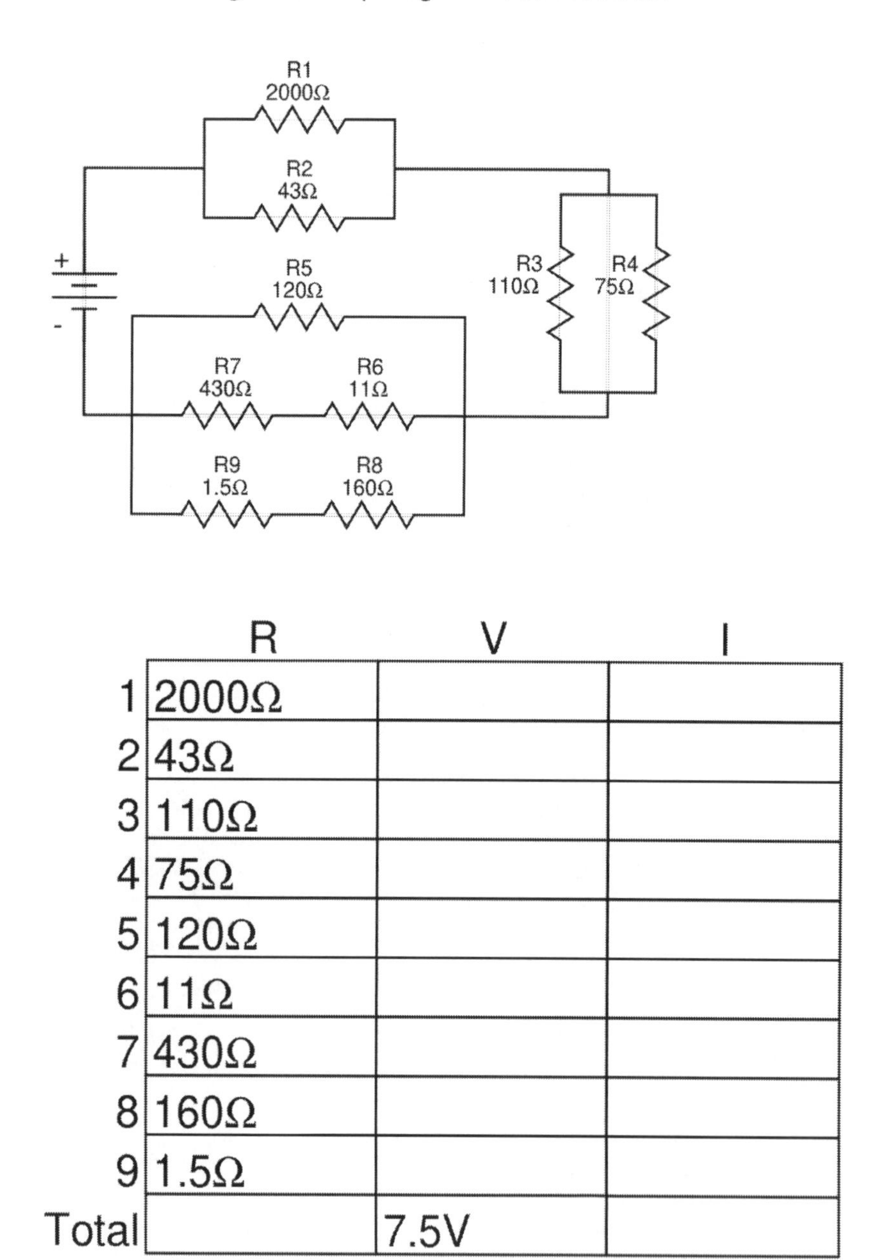

	R	V	I
1	2000Ω		
2	43Ω		
3	110Ω		
4	75Ω		
5	120Ω		
6	11Ω		
7	430Ω		
8	160Ω		
9	1.5Ω		
Total		7.5V	

Exercise #44

Find the voltage and amperage for each resistor.

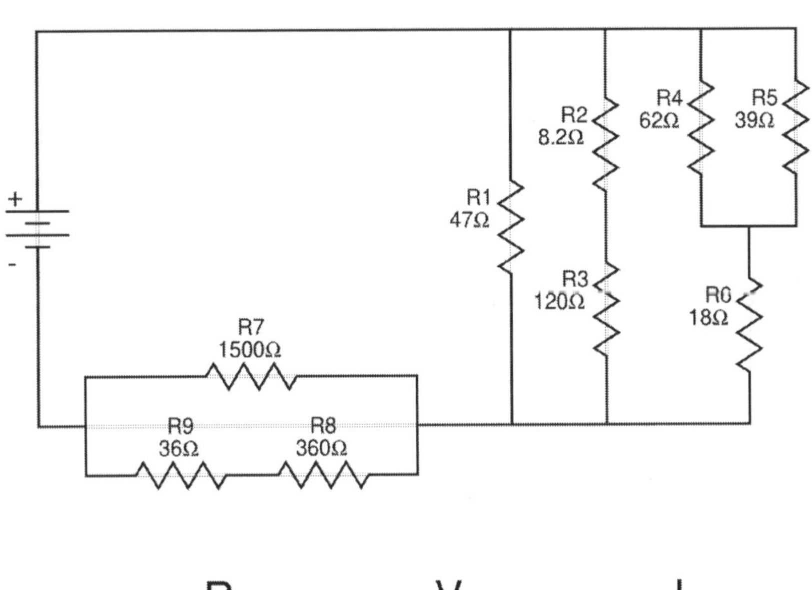

	R	V	I
1	47Ω		
2	8.2Ω		
3	120Ω		
4	62Ω		
5	39Ω		
6	18Ω		
7	1500Ω		
8	360Ω		
9	36Ω		
Total		18V	

Exercise #45

Find the voltage and amperage for each resistor.

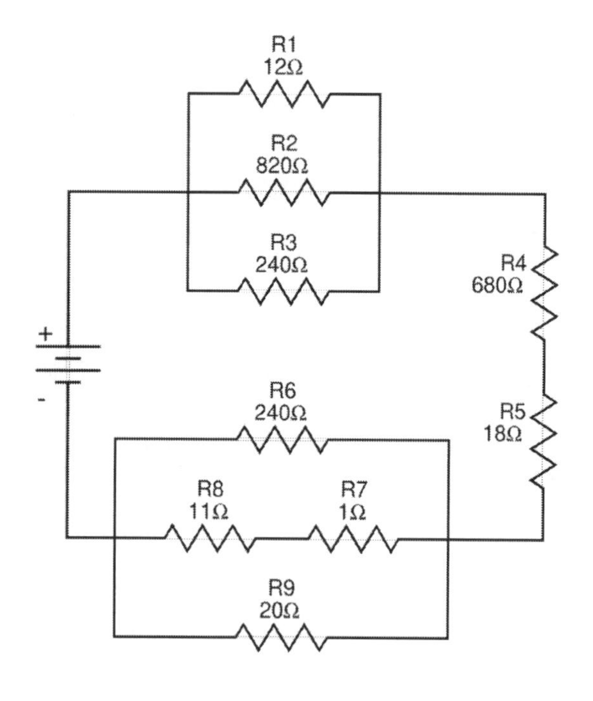

	R	V	I
1	12Ω		
2	820Ω		
3	240Ω		
4	680Ω		
5	18Ω		
6	240Ω		
7	1Ω		
8	11Ω		
9	20Ω		
Total		240V	

Exercise #46

Find the voltage and amperage for each resistor.

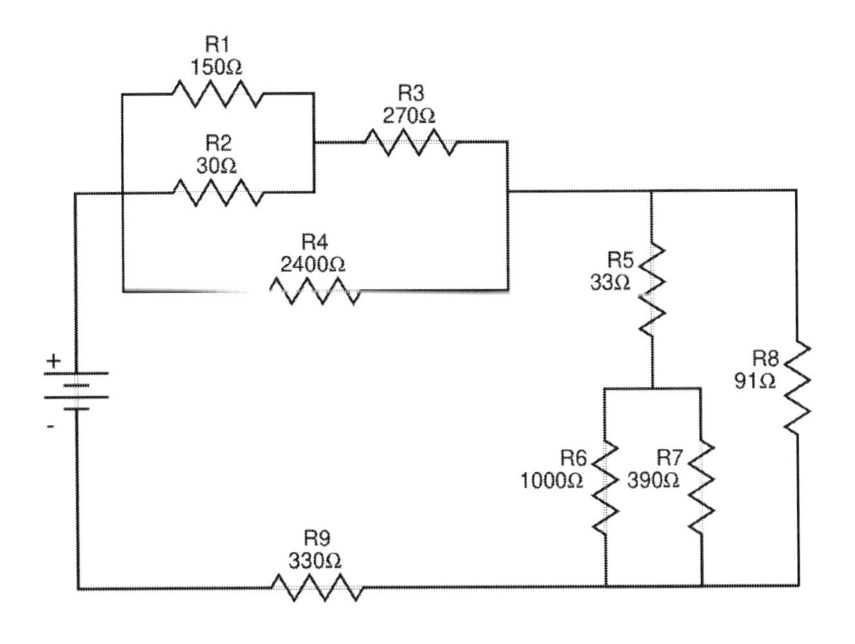

	R	V	I
1	150Ω		
2	30Ω		
3	270Ω		
4	2400Ω		
5	33Ω		
6	1000Ω		
7	390Ω		
8	91Ω		
9	330Ω		
Total		9V	

Exercise #47

Find the voltage and amperage for each resistor.

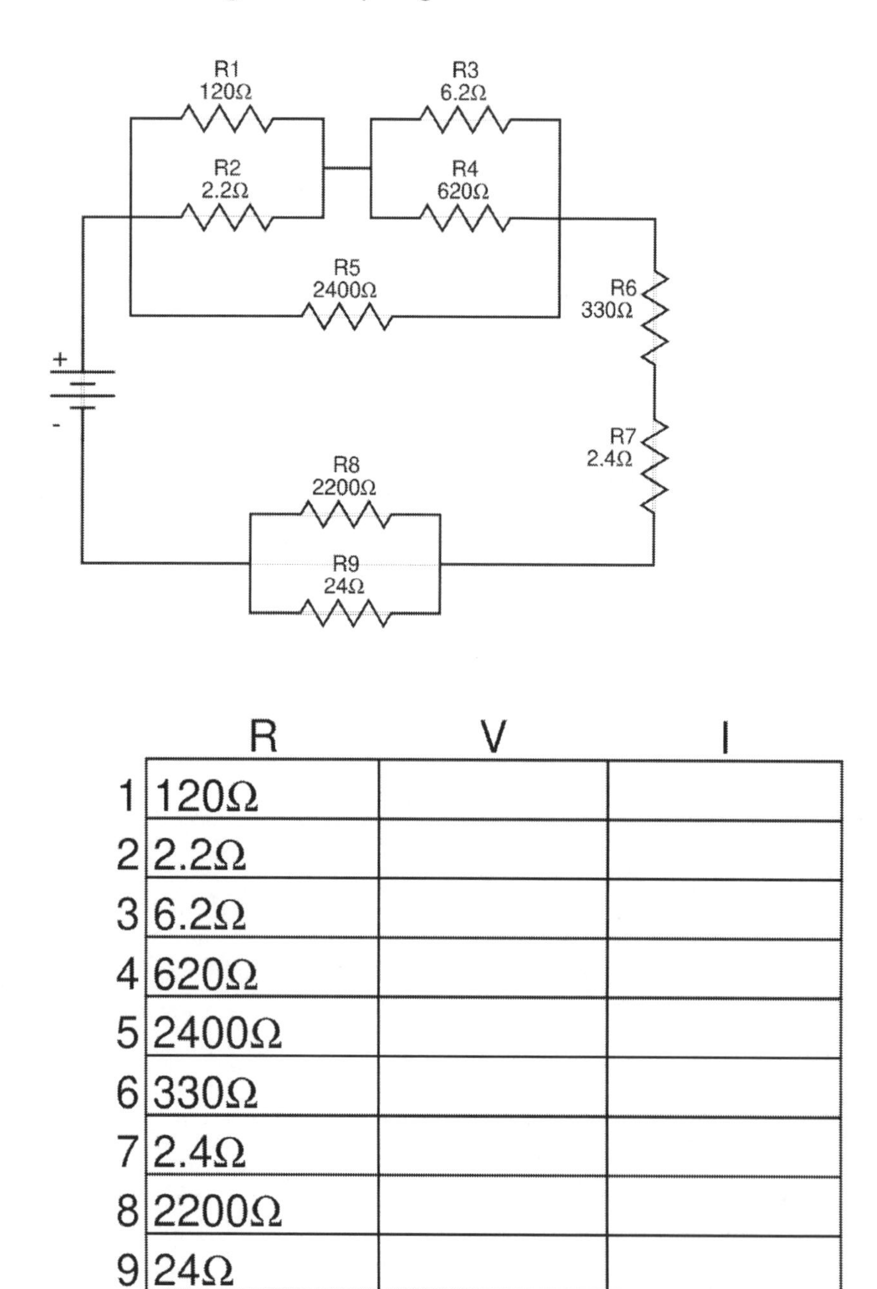

	R	V	I
1	120Ω		
2	2.2Ω		
3	6.2Ω		
4	620Ω		
5	2400Ω		
6	330Ω		
7	2.4Ω		
8	2200Ω		
9	24Ω		
Total		10.5V	

Exercise #48

Find the voltage and amperage for each resistor.

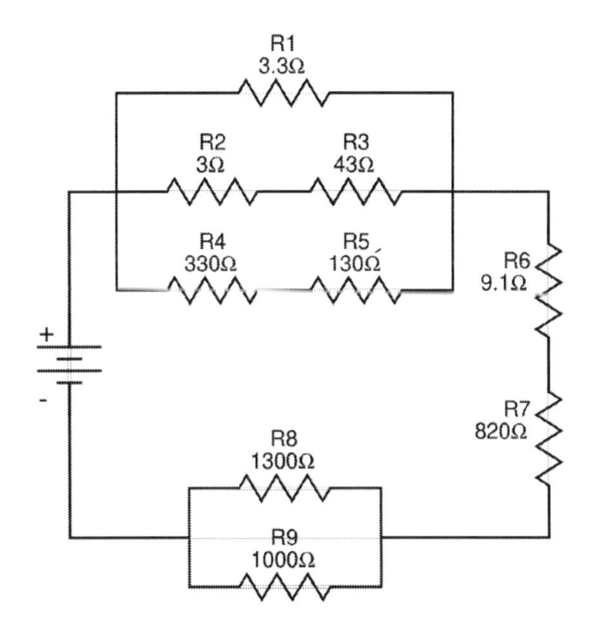

	R	V	I
1	3.3Ω		
2	3Ω		
3	43Ω		
4	330Ω		
5	130Ω		
6	9.1Ω		
7	820Ω		
8	1300Ω		
9	1000Ω		
Total		30V	

Exercise #49

Find the voltage and amperage for each resistor.

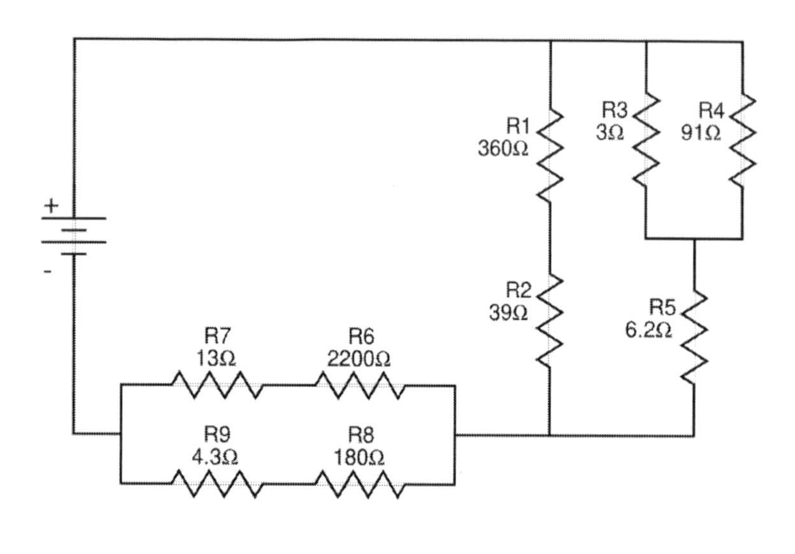

	R	V	I
1	360Ω		
2	39Ω		
3	3Ω		
4	91Ω		
5	6.2Ω		
6	2200Ω		
7	13Ω		
8	180Ω		
9	4.3Ω		
Total		36V	

Exercise #50

Find the voltage and amperage for each resistor.

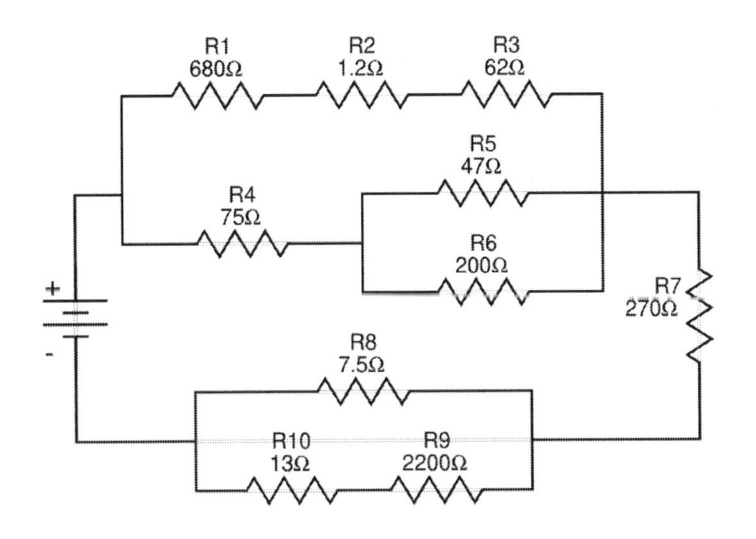

	R	V	I
1	680Ω		
2	1.2Ω		
3	62Ω		
4	75Ω		
5	47Ω		
6	200Ω		
7	270Ω		
8	7.5Ω		
9	2200Ω		
10	13Ω		
Total		225V	

Exercise #51

Find the voltage and amperage for each resistor.

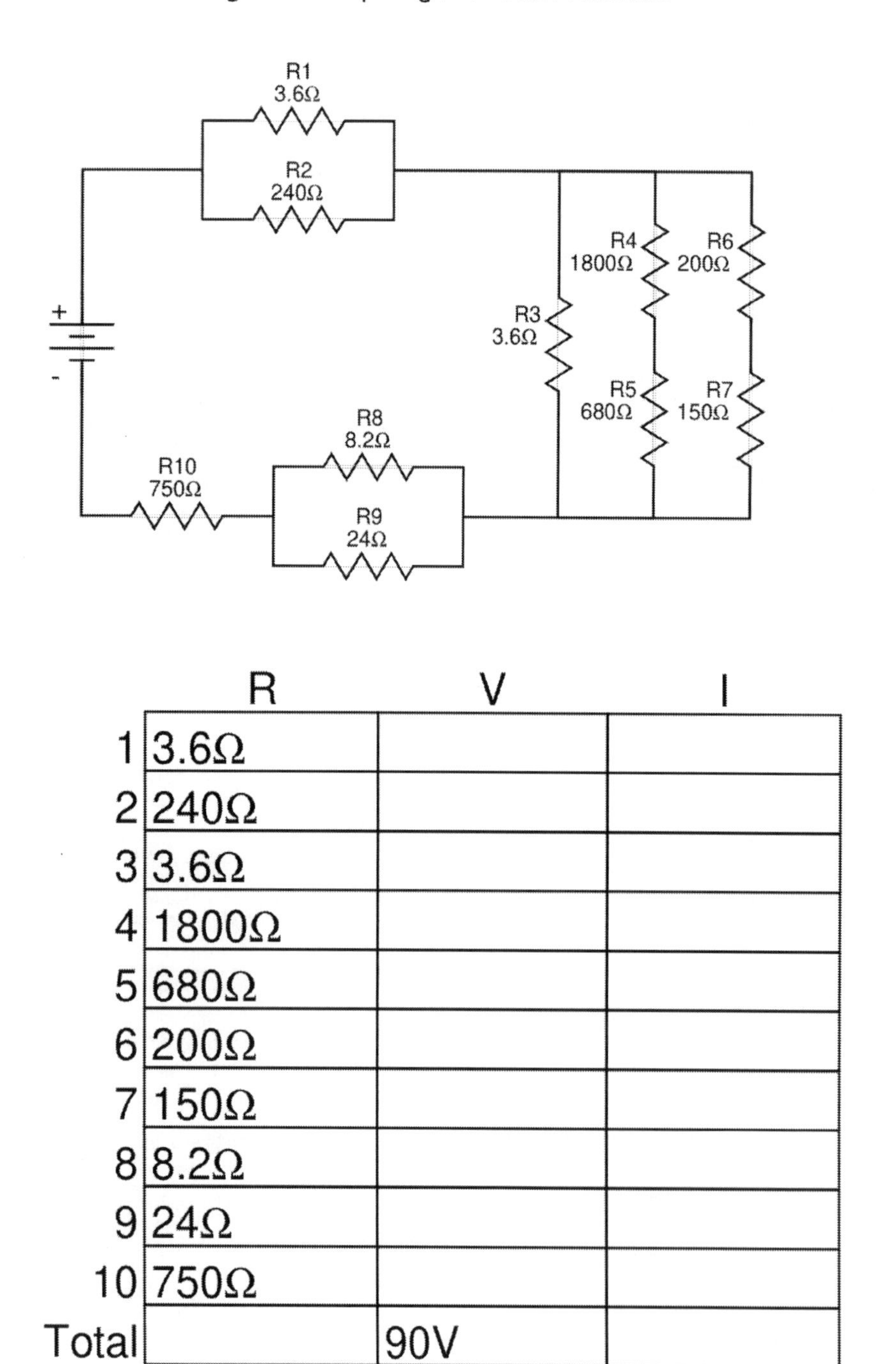

	R	V	I
1	3.6Ω		
2	240Ω		
3	3.6Ω		
4	1800Ω		
5	680Ω		
6	200Ω		
7	150Ω		
8	8.2Ω		
9	24Ω		
10	750Ω		
Total		90V	

Exercise #52

Find the voltage and amperage for each resistor.

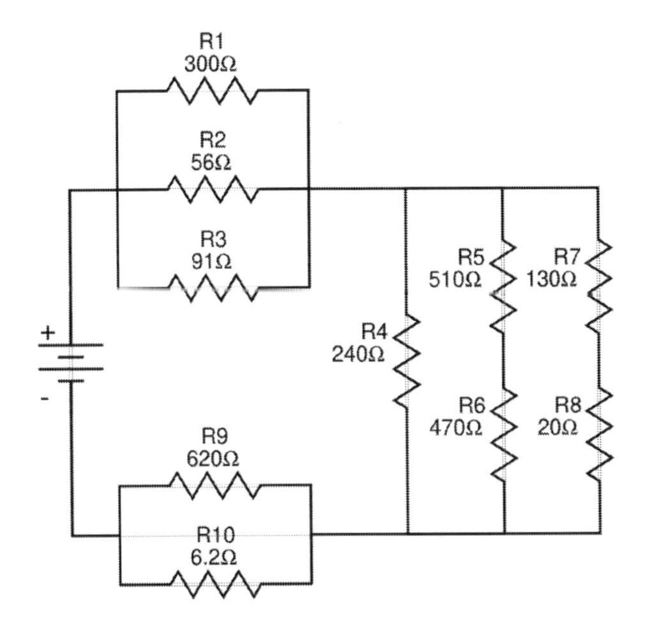

	R	V	I
1	300Ω		
2	56Ω		
3	91Ω		
4	240Ω		
5	510Ω		
6	470Ω		
7	130Ω		
8	20Ω		
9	620Ω		
10	6.2Ω		
Total		195V	

Exercise #53

Find the voltage and amperage for each resistor.

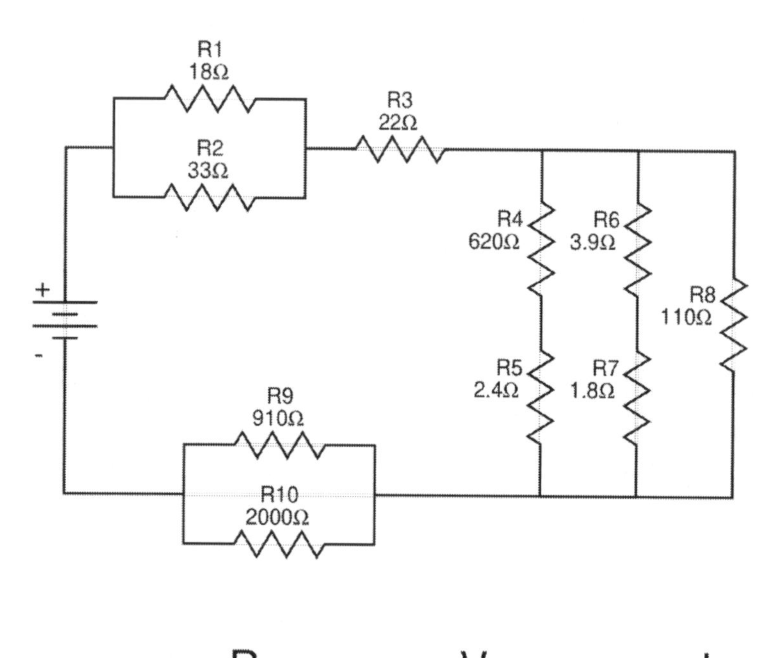

	R	V	I
1	18Ω		
2	33Ω		
3	22Ω		
4	620Ω		
5	2.4Ω		
6	3.9Ω		
7	1.8Ω		
8	110Ω		
9	910Ω		
10	2000Ω		
Total		135V	

Exercise #54

Find the voltage and amperage for each resistor.

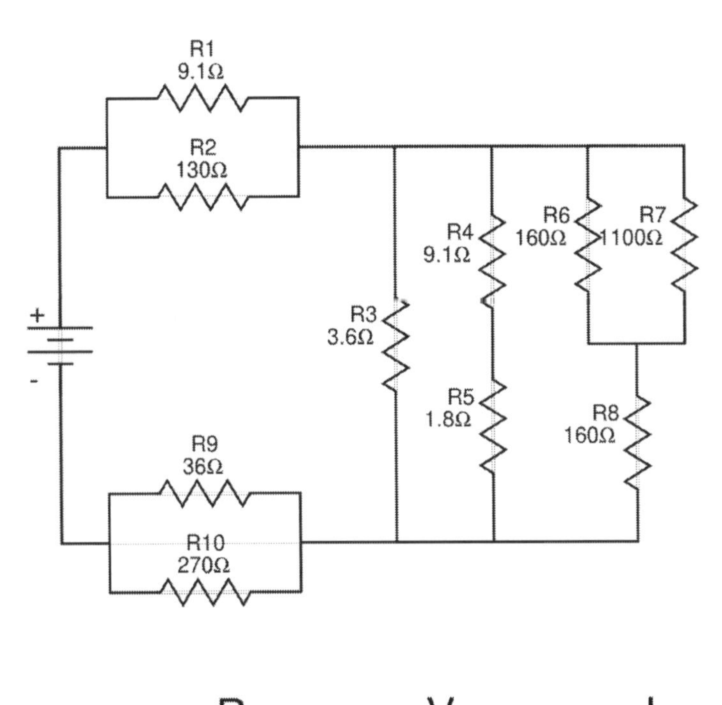

	R	V	I
1	9.1Ω		
2	130Ω		
3	3.6Ω		
4	9.1Ω		
5	1.8Ω		
6	160Ω		
7	1100Ω		
8	160Ω		
9	36Ω		
10	270Ω		
Total		66V	

Exercise #55

Find the voltage and amperage for each resistor.

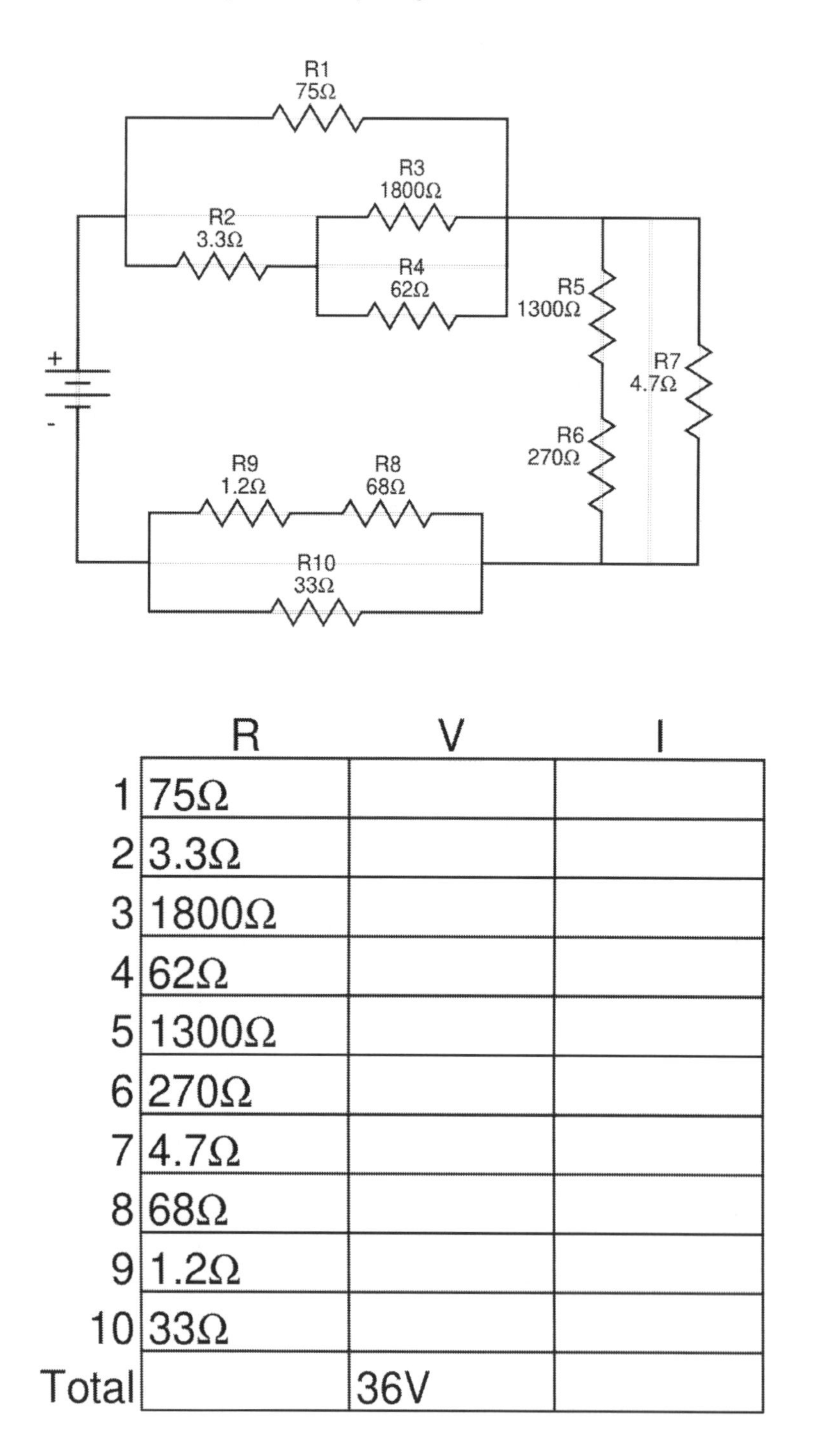

	R	V	I
1	75Ω		
2	3.3Ω		
3	1800Ω		
4	62Ω		
5	1300Ω		
6	270Ω		
7	4.7Ω		
8	68Ω		
9	1.2Ω		
10	33Ω		
Total		36V	

Exercise #56

Find the voltage and amperage for each resistor.

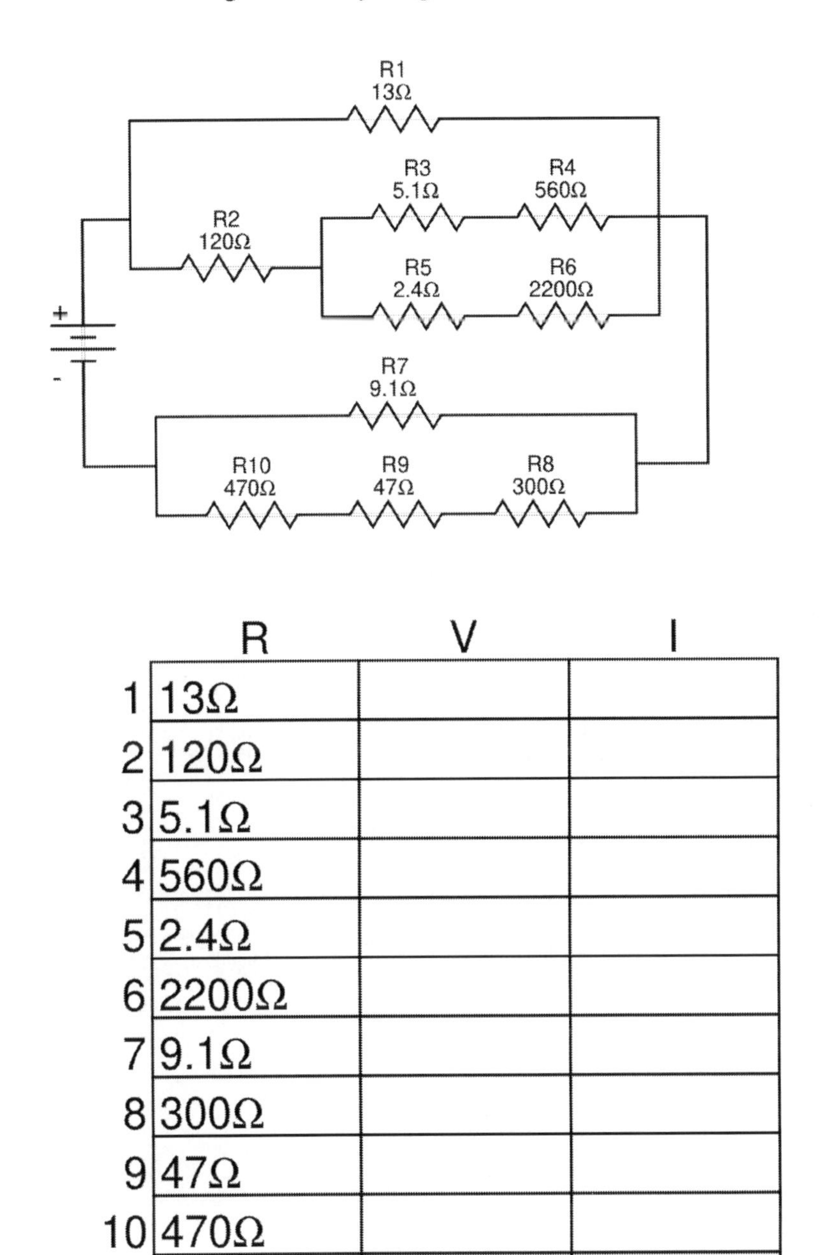

	R	V	I
1	13Ω		
2	120Ω		
3	5.1Ω		
4	560Ω		
5	2.4Ω		
6	2200Ω		
7	9.1Ω		
8	300Ω		
9	47Ω		
10	470Ω		
Total		60V	

Exercise #57

Find the voltage and amperage for each resistor.

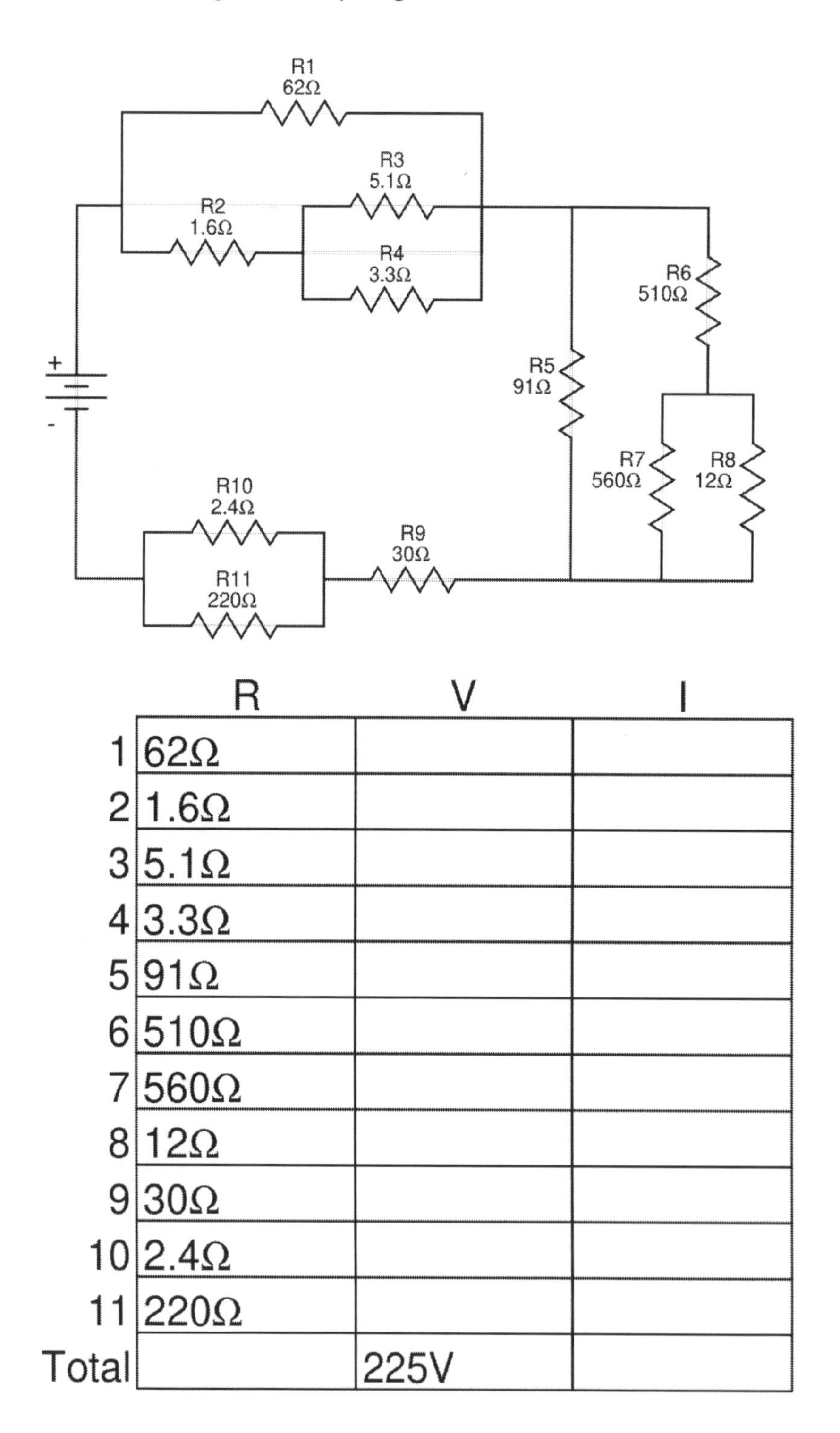

	R	V	I
1	62Ω		
2	1.6Ω		
3	5.1Ω		
4	3.3Ω		
5	91Ω		
6	510Ω		
7	560Ω		
8	12Ω		
9	30Ω		
10	2.4Ω		
11	220Ω		
Total		225V	

Exercise #58

Find the voltage and amperage for each resistor.

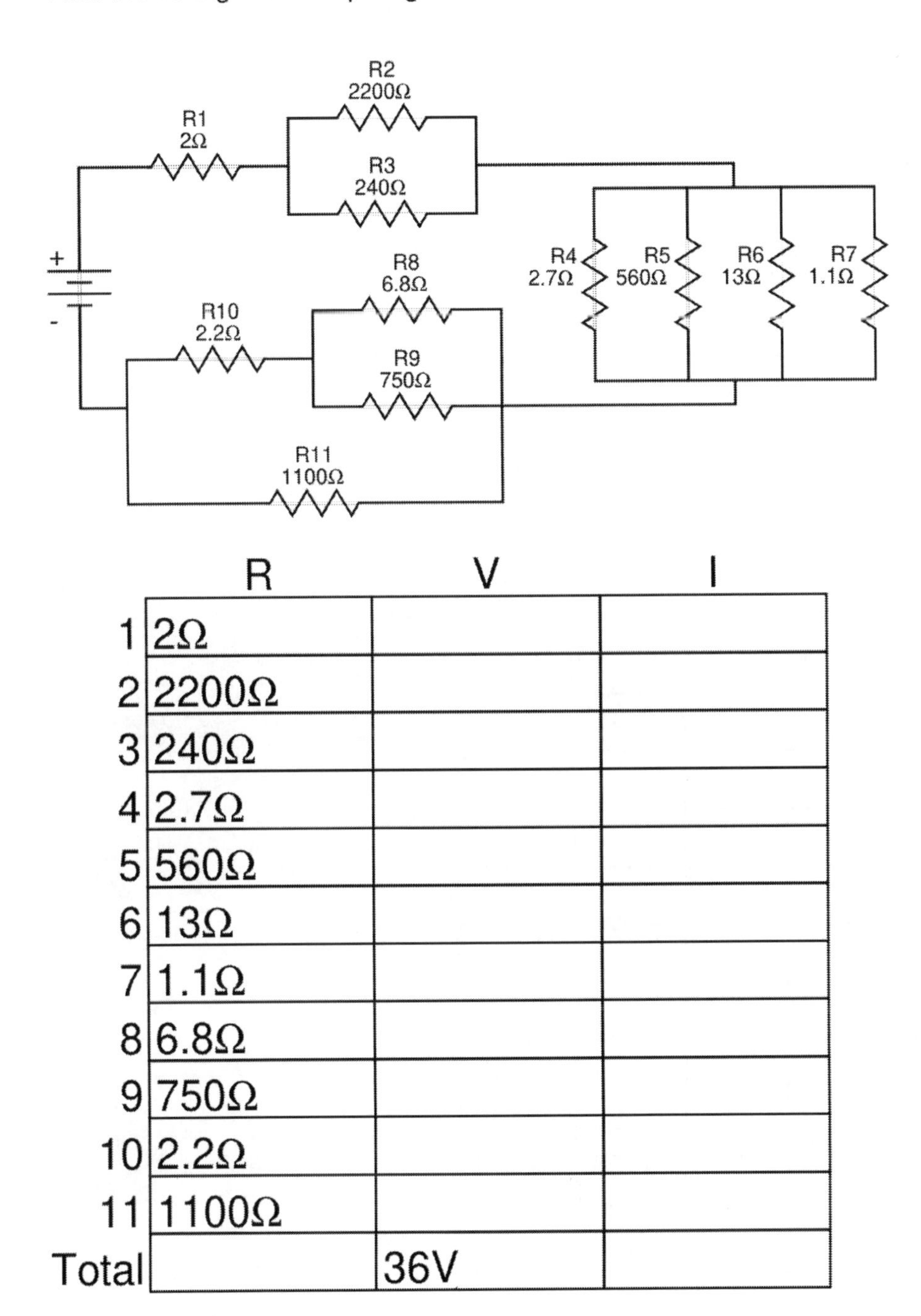

	R	V	I
1	2Ω		
2	2200Ω		
3	240Ω		
4	2.7Ω		
5	560Ω		
6	13Ω		
7	1.1Ω		
8	6.8Ω		
9	750Ω		
10	2.2Ω		
11	1100Ω		
Total		36V	

Exercise #59

Find the voltage and amperage for each resistor.

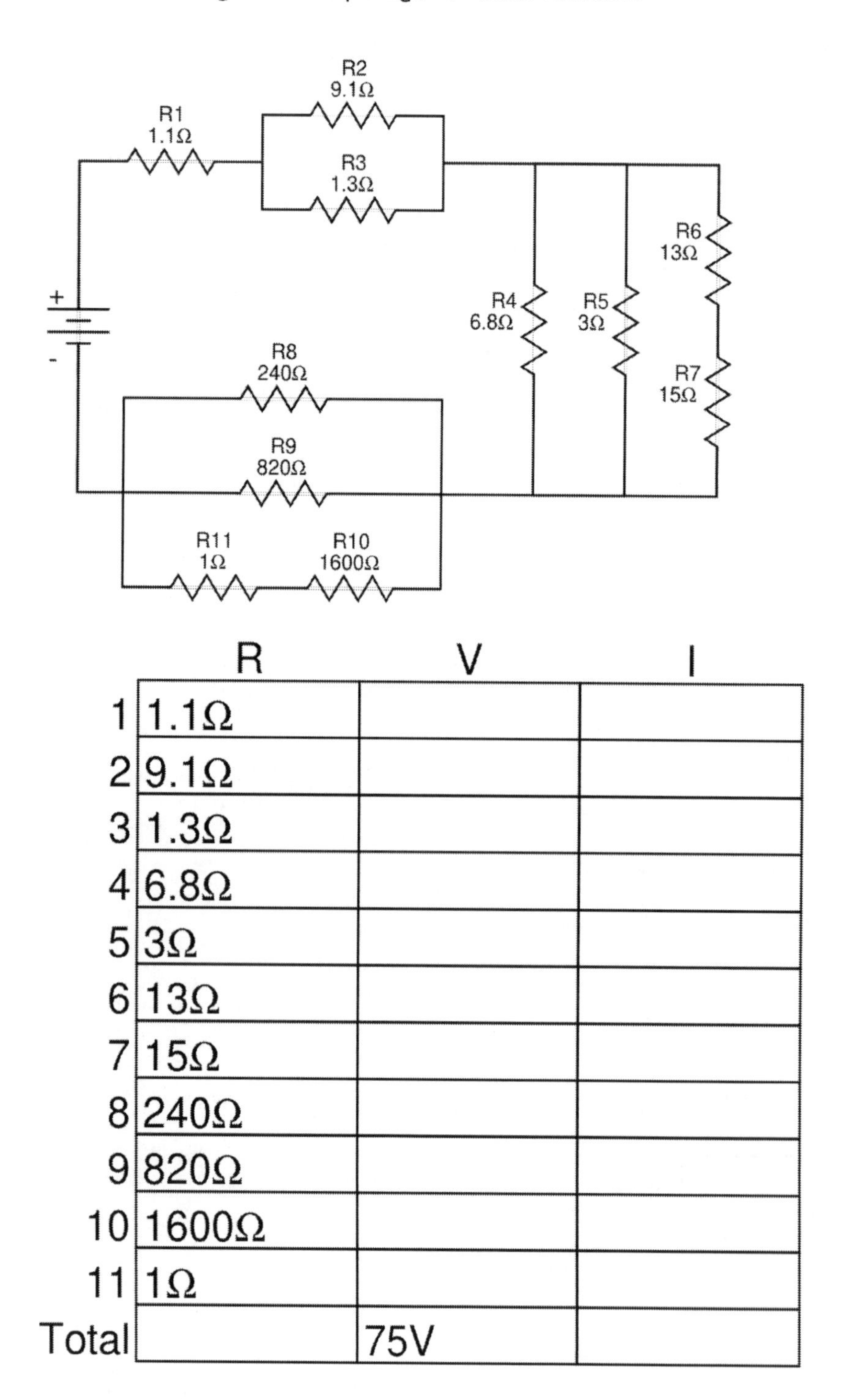

	R	V	I
1	1.1Ω		
2	9.1Ω		
3	1.3Ω		
4	6.8Ω		
5	3Ω		
6	13Ω		
7	15Ω		
8	240Ω		
9	820Ω		
10	1600Ω		
11	1Ω		
Total		75V	

Exercise #60

Find the voltage and amperage for each resistor.

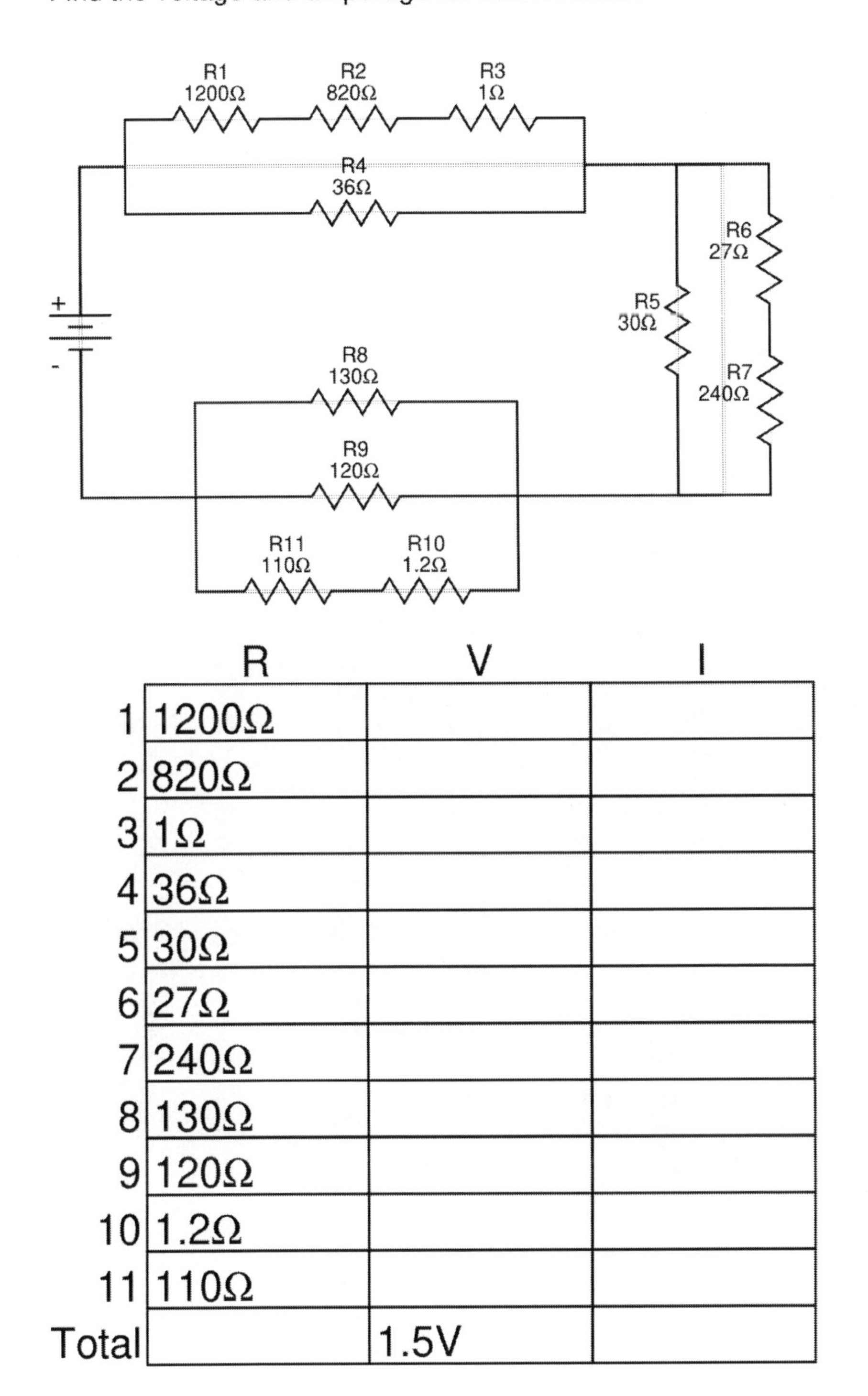

	R	V	I
1	1200Ω		
2	820Ω		
3	1Ω		
4	36Ω		
5	30Ω		
6	27Ω		
7	240Ω		
8	130Ω		
9	120Ω		
10	1.2Ω		
11	110Ω		
Total		1.5V	

Exercise #61

Find the voltage and amperage for each resistor.

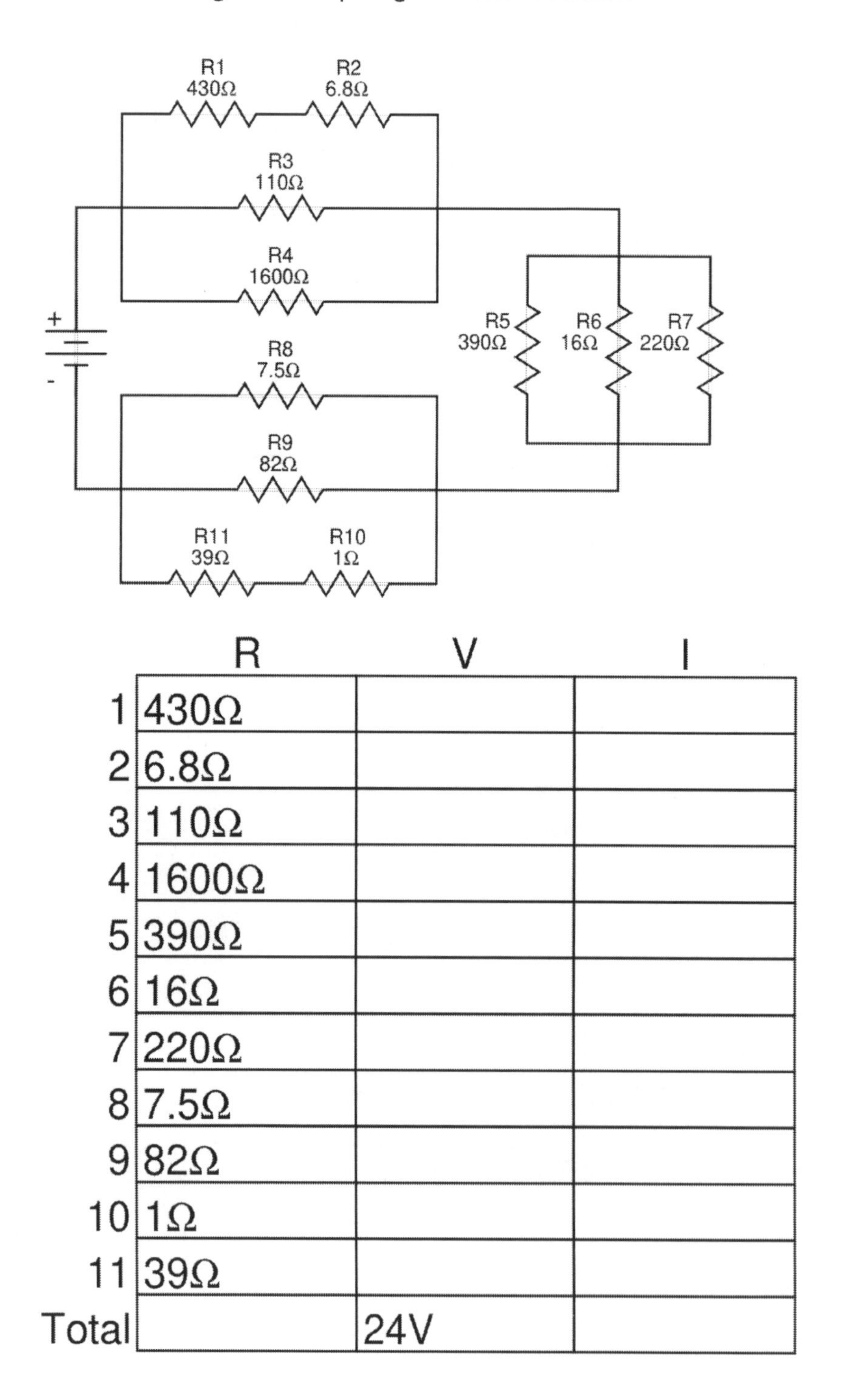

	R	V	I
1	430Ω		
2	6.8Ω		
3	110Ω		
4	1600Ω		
5	390Ω		
6	16Ω		
7	220Ω		
8	7.5Ω		
9	82Ω		
10	1Ω		
11	39Ω		
Total		24V	

Exercise #62

Find the voltage and amperage for each resistor.

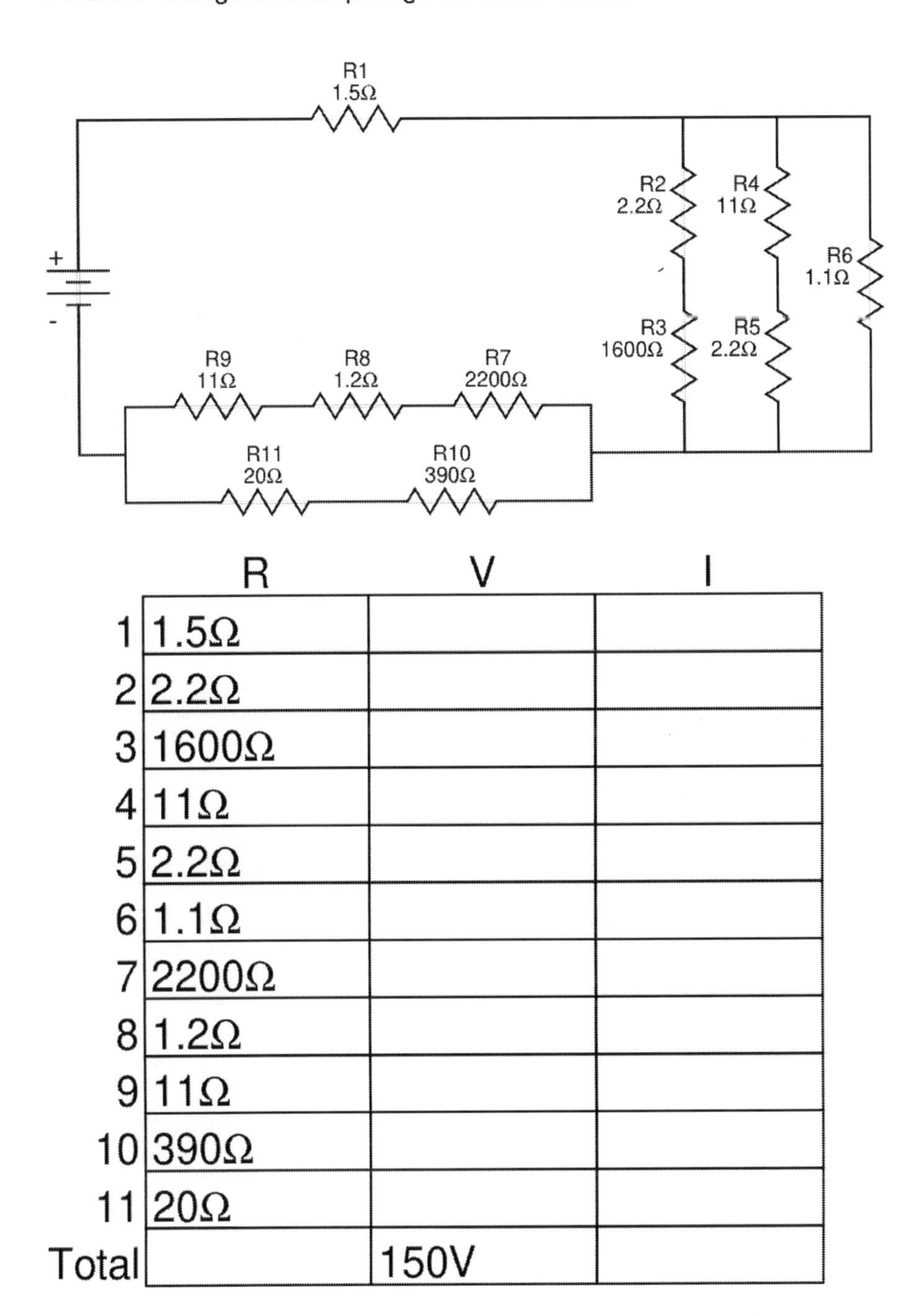

	R	V	I
1	1.5Ω		
2	2.2Ω		
3	1600Ω		
4	11Ω		
5	2.2Ω		
6	1.1Ω		
7	2200Ω		
8	1.2Ω		
9	11Ω		
10	390Ω		
11	20Ω		
Total		150V	

Exercise #63

Find the voltage and amperage for each resistor.

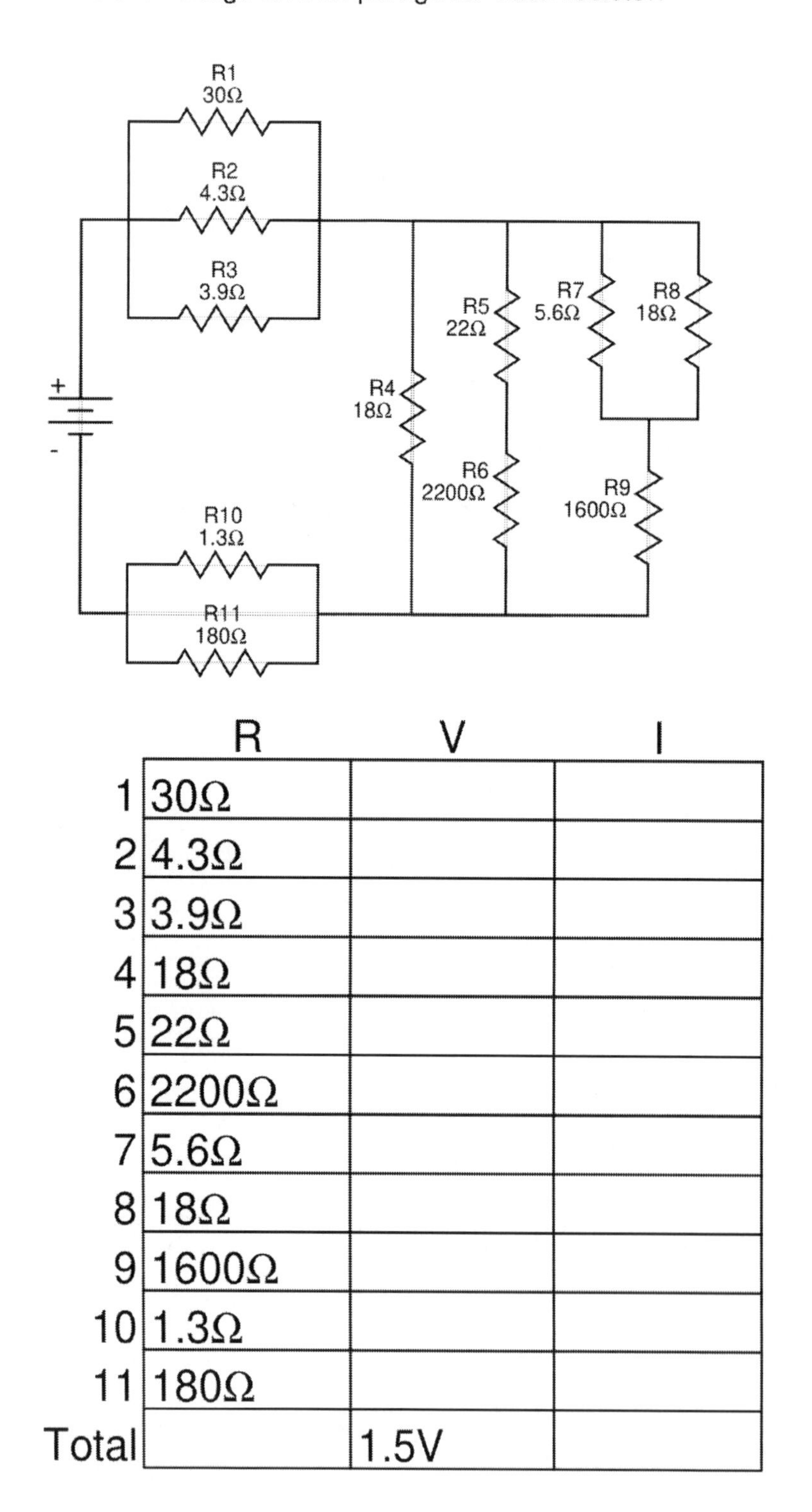

	R	V	I
1	30Ω		
2	4.3Ω		
3	3.9Ω		
4	18Ω		
5	22Ω		
6	2200Ω		
7	5.6Ω		
8	18Ω		
9	1600Ω		
10	1.3Ω		
11	180Ω		
Total		1.5V	

Exercise #64

Find the voltage and amperage for each resistor.

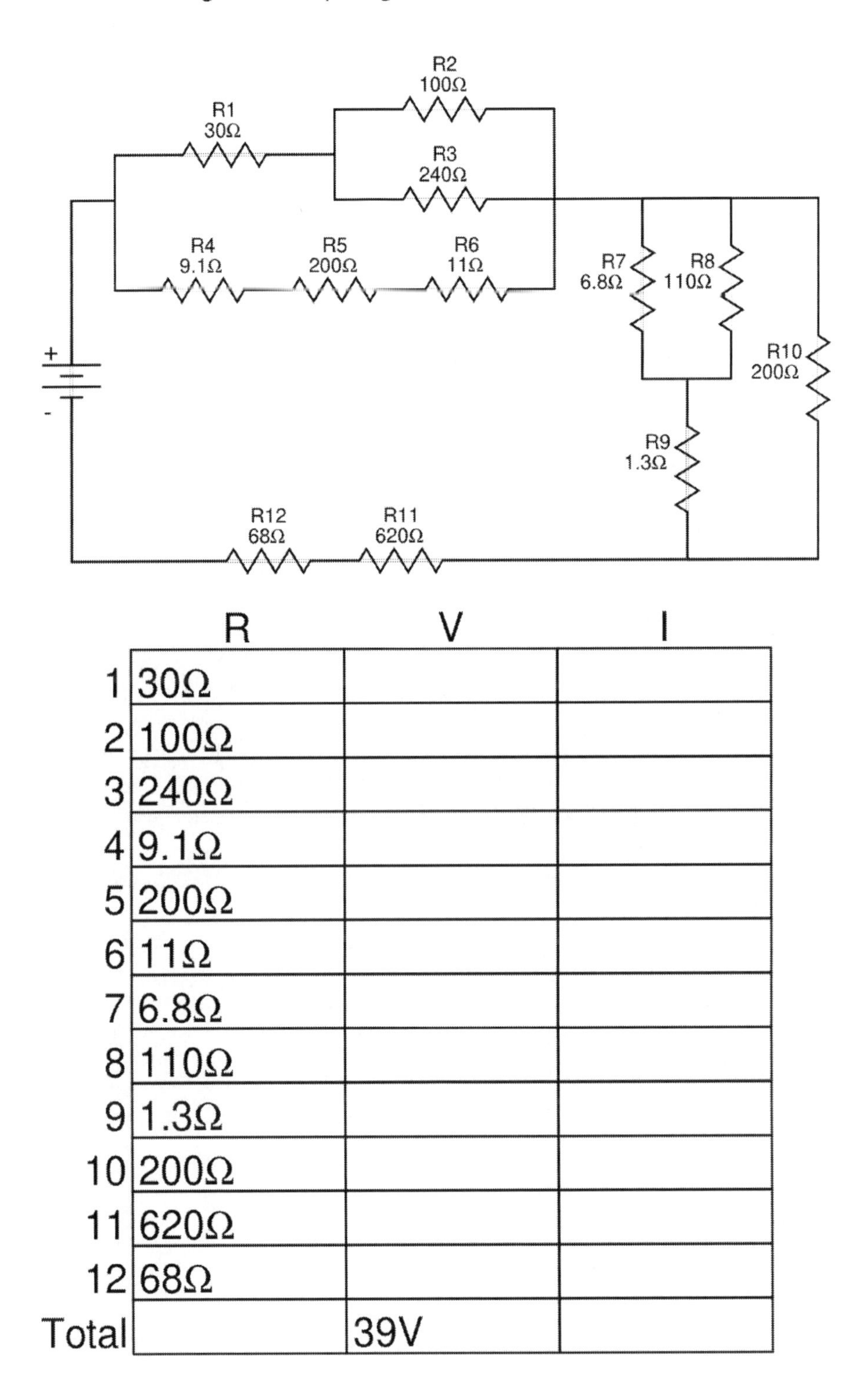

	R	V	I
1	30Ω		
2	100Ω		
3	240Ω		
4	9.1Ω		
5	200Ω		
6	11Ω		
7	6.8Ω		
8	110Ω		
9	1.3Ω		
10	200Ω		
11	620Ω		
12	68Ω		
Total		39V	

Exercise #65

Find the voltage and amperage for each resistor.

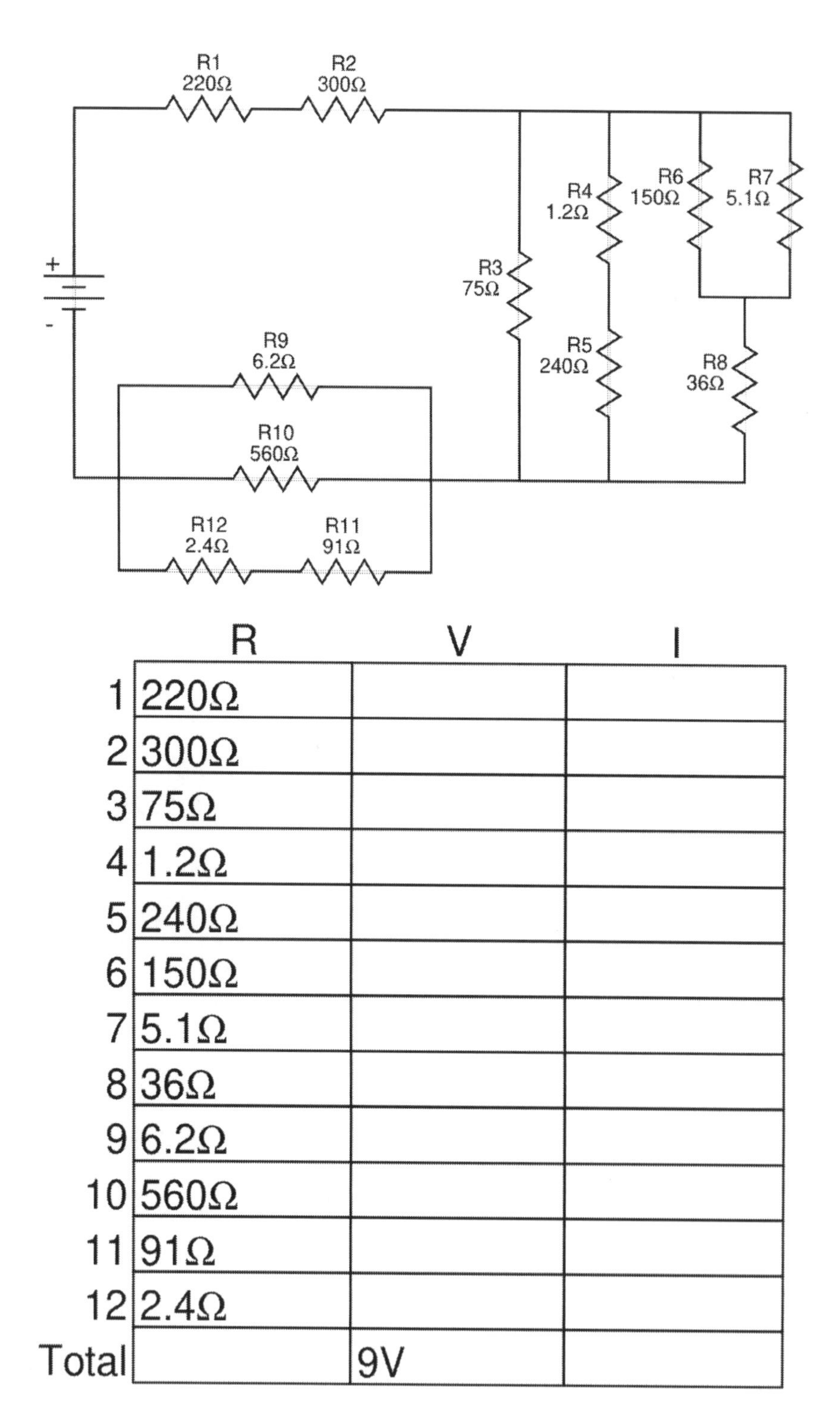

	R	V	I
1	220Ω		
2	300Ω		
3	75Ω		
4	1.2Ω		
5	240Ω		
6	150Ω		
7	5.1Ω		
8	36Ω		
9	6.2Ω		
10	560Ω		
11	91Ω		
12	2.4Ω		
Total		9V	

Exercise #66

Find the voltage and amperage for each resistor.

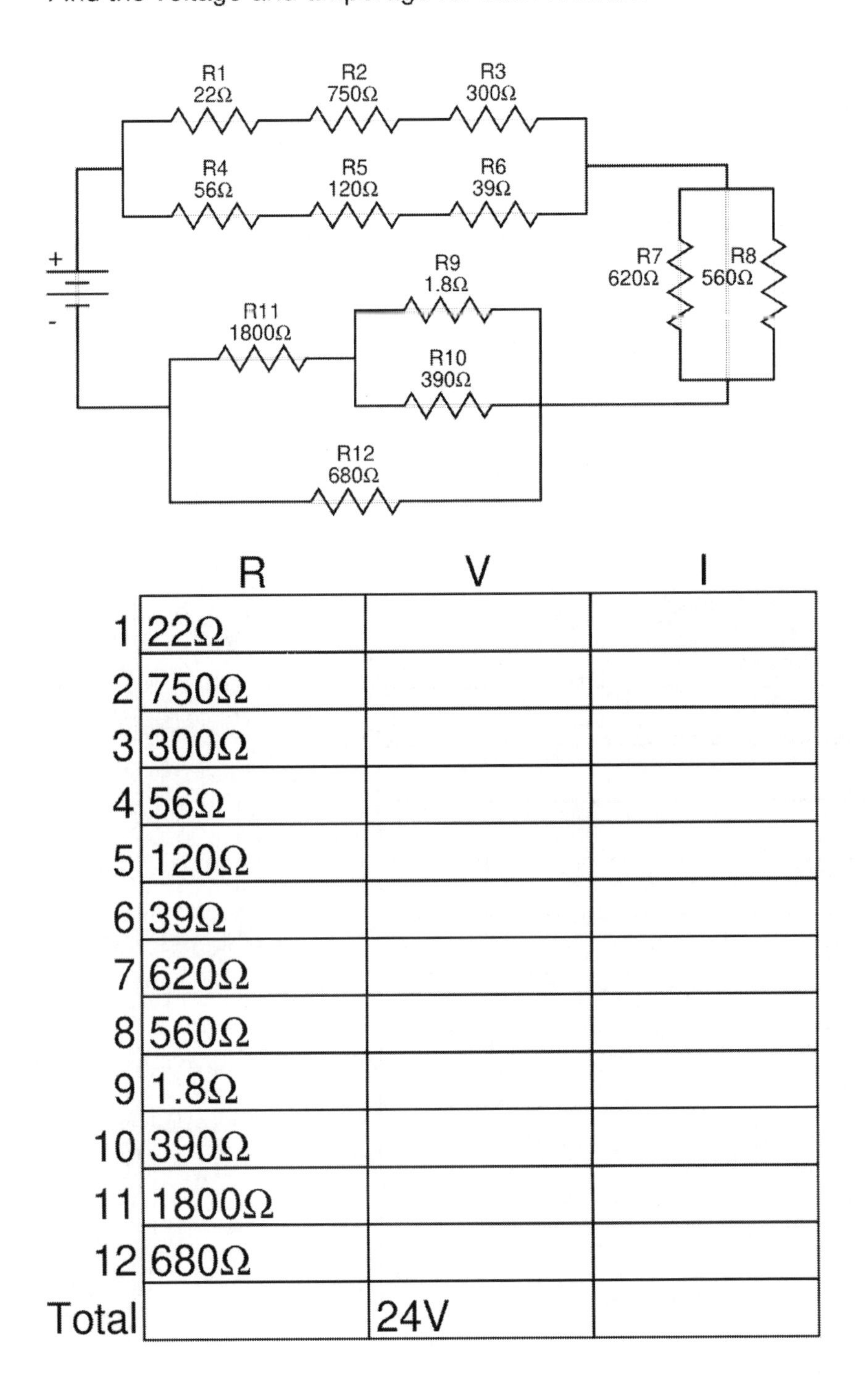

	R	V	I
1	22Ω		
2	750Ω		
3	300Ω		
4	56Ω		
5	120Ω		
6	39Ω		
7	620Ω		
8	560Ω		
9	1.8Ω		
10	390Ω		
11	1800Ω		
12	680Ω		
Total		24V	

Exercise #67

Find the voltage and amperage for each resistor.

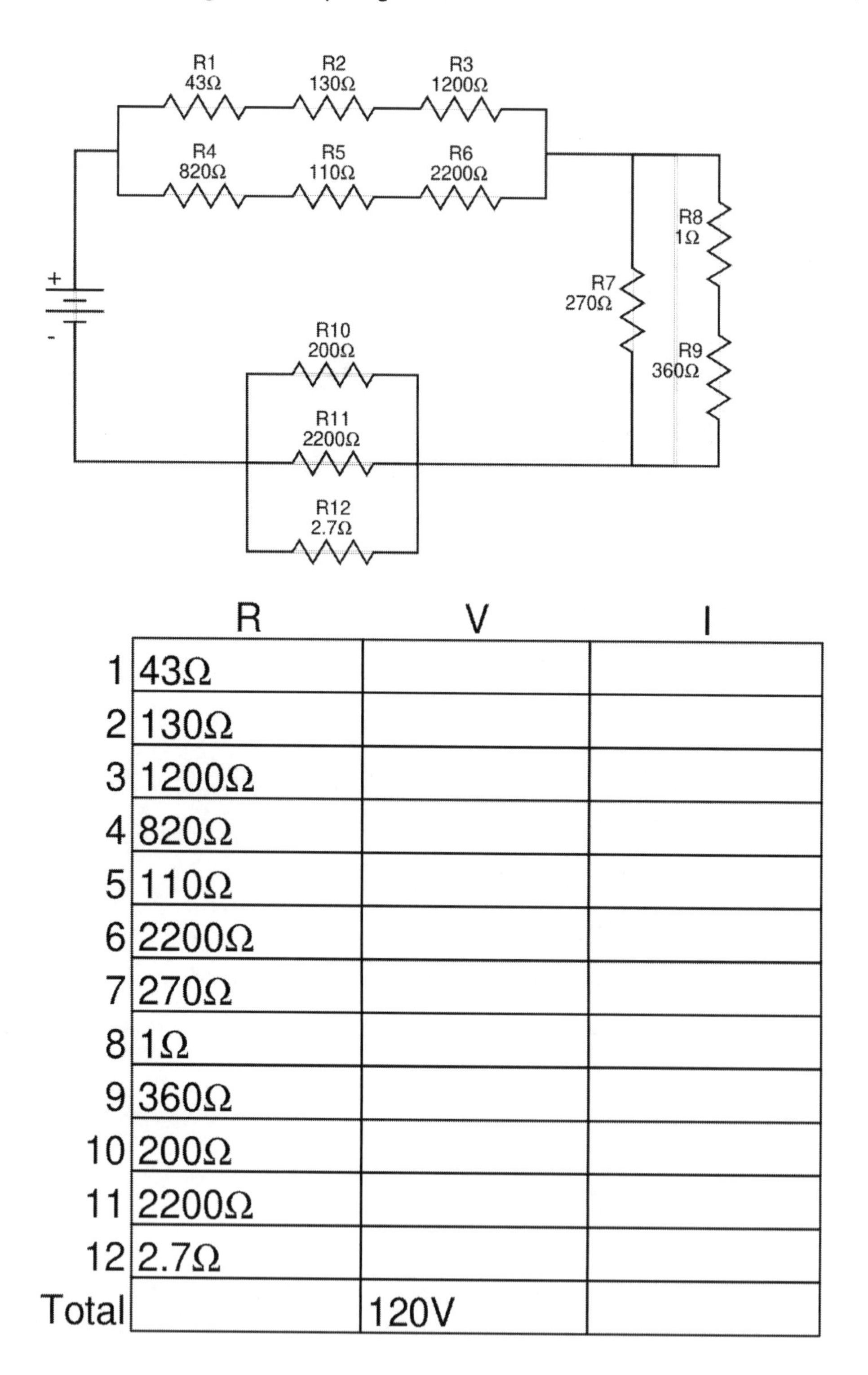

	R	V	I
1	43Ω		
2	130Ω		
3	1200Ω		
4	820Ω		
5	110Ω		
6	2200Ω		
7	270Ω		
8	1Ω		
9	360Ω		
10	200Ω		
11	2200Ω		
12	2.7Ω		
Total		120V	

Exercise #68

Find the voltage and amperage for each resistor.

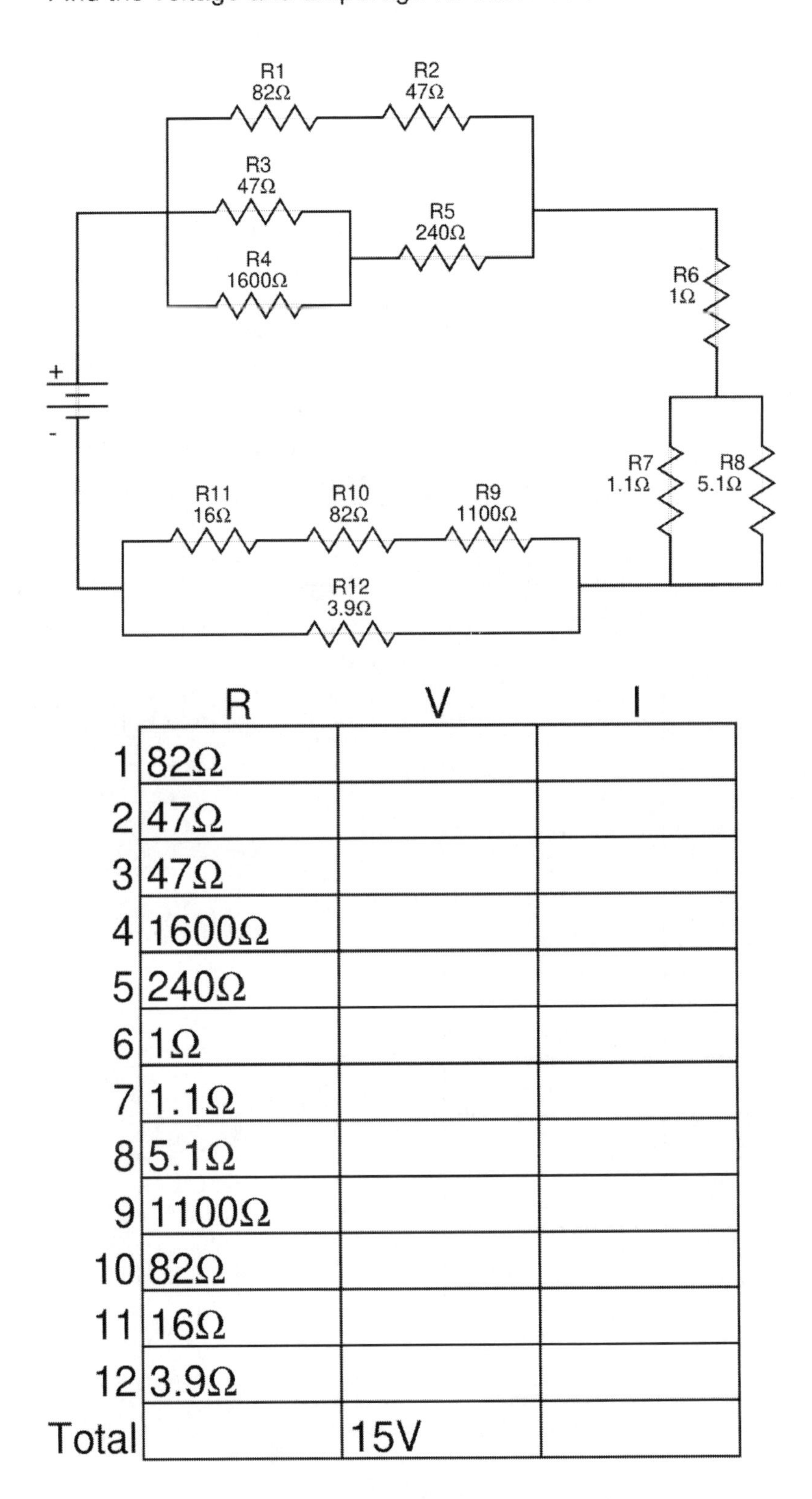

	R	V	I
1	82Ω		
2	47Ω		
3	47Ω		
4	1600Ω		
5	240Ω		
6	1Ω		
7	1.1Ω		
8	5.1Ω		
9	1100Ω		
10	82Ω		
11	16Ω		
12	3.9Ω		
Total		15V	

Exercise #69

Find the voltage and amperage for each resistor.

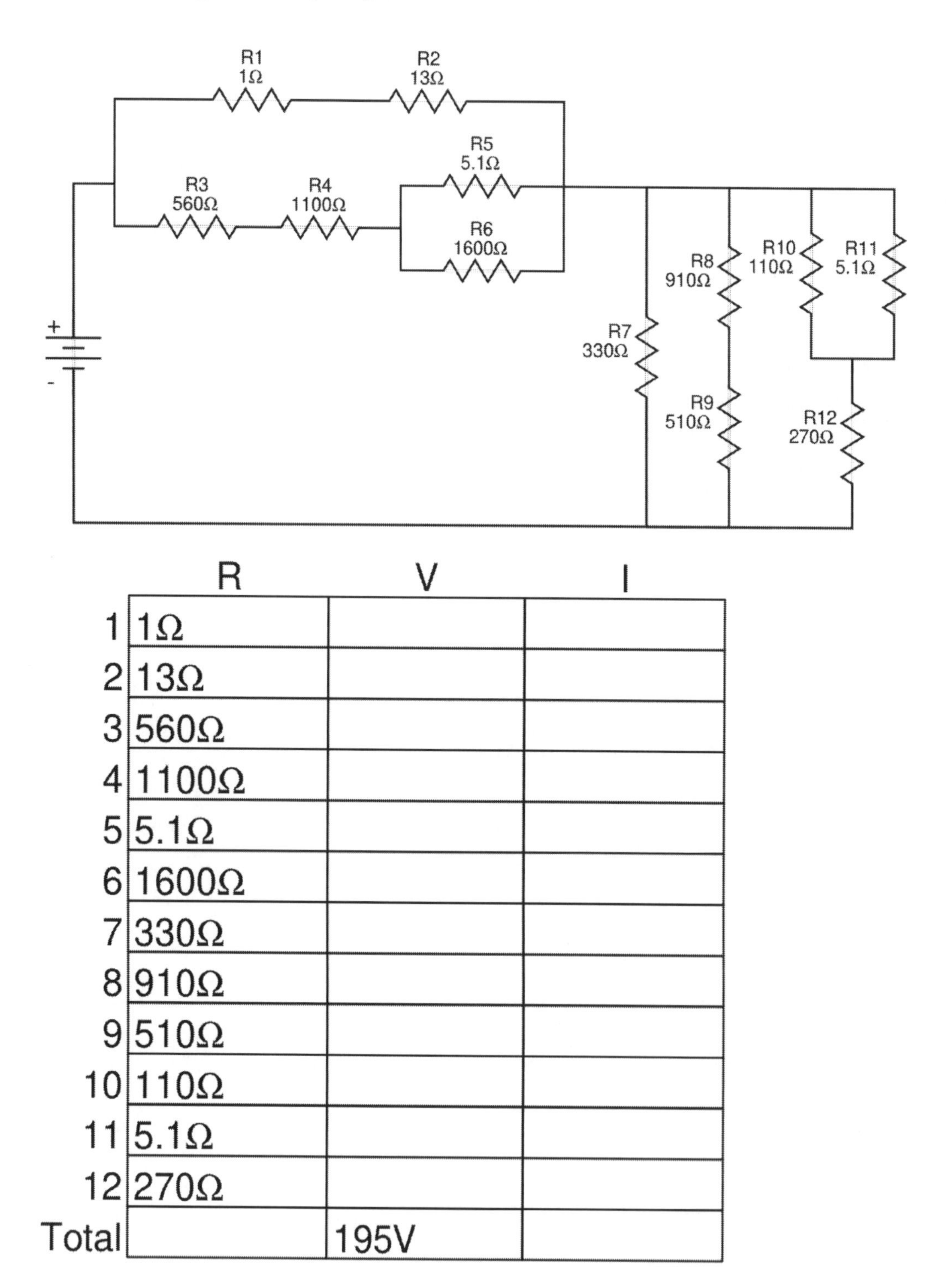

	R	V	I
1	1Ω		
2	13Ω		
3	560Ω		
4	1100Ω		
5	5.1Ω		
6	1600Ω		
7	330Ω		
8	910Ω		
9	510Ω		
10	110Ω		
11	5.1Ω		
12	270Ω		
Total		195V	

Exercise #70

Find the voltage and amperage for each resistor.

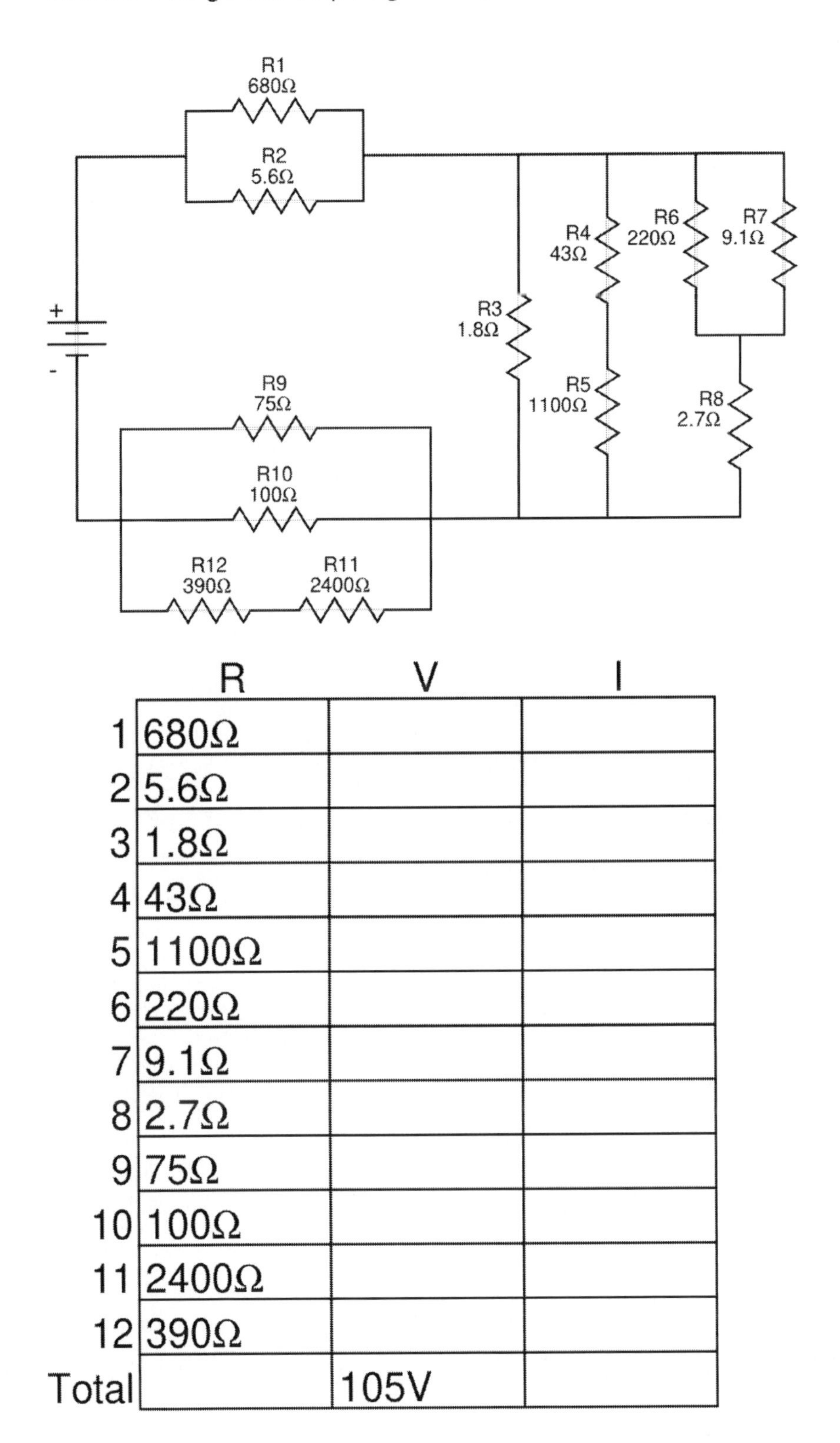

	R	V	I
1	680Ω		
2	5.6Ω		
3	1.8Ω		
4	43Ω		
5	1100Ω		
6	220Ω		
7	9.1Ω		
8	2.7Ω		
9	75Ω		
10	100Ω		
11	2400Ω		
12	390Ω		
Total		105V	

Exercise #71

Find the voltage and amperage for each resistor.

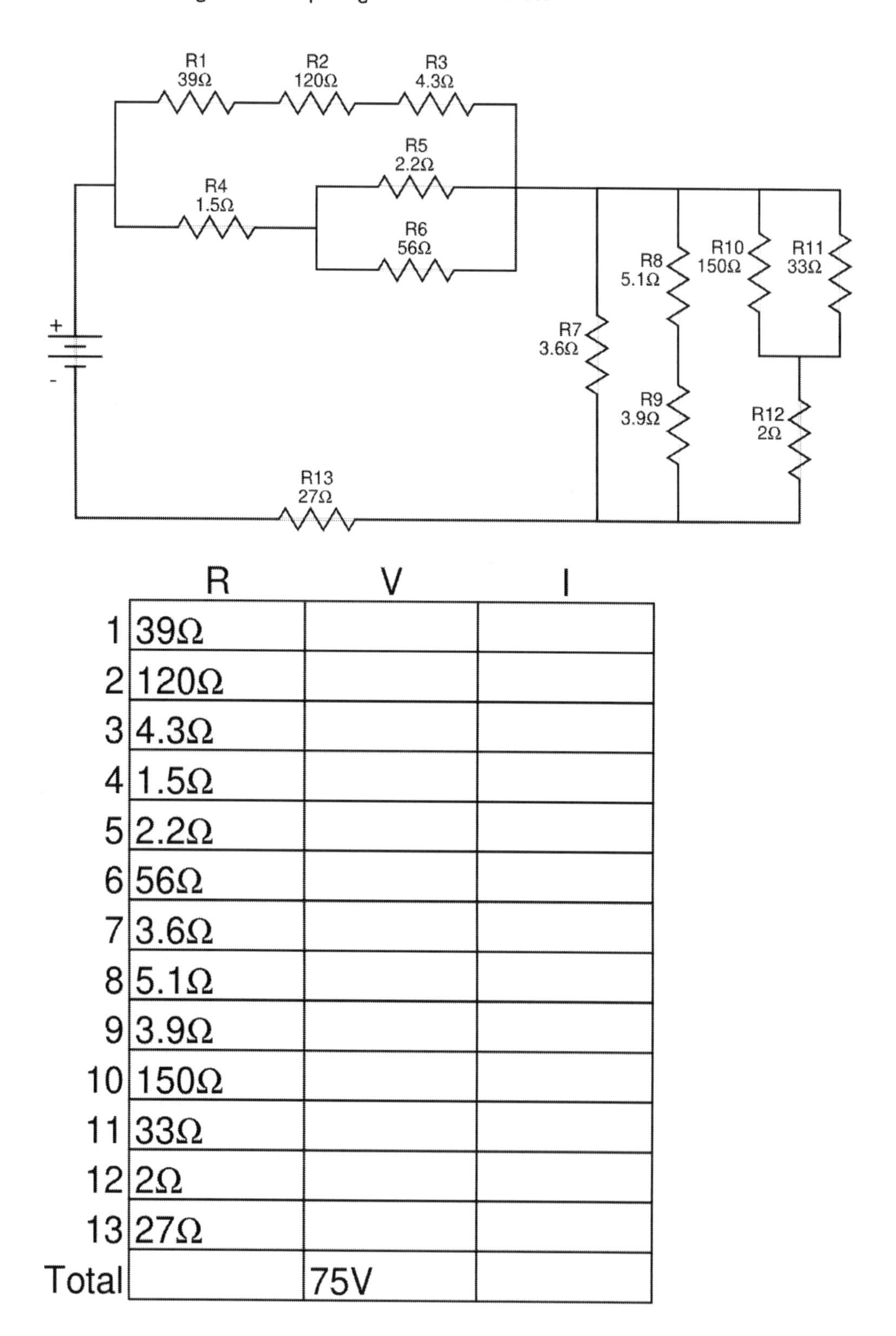

	R	V	I
1	39Ω		
2	120Ω		
3	4.3Ω		
4	1.5Ω		
5	2.2Ω		
6	56Ω		
7	3.6Ω		
8	5.1Ω		
9	3.9Ω		
10	150Ω		
11	33Ω		
12	2Ω		
13	27Ω		
Total		75V	

Exercise #72

Find the voltage and amperage for each resistor.

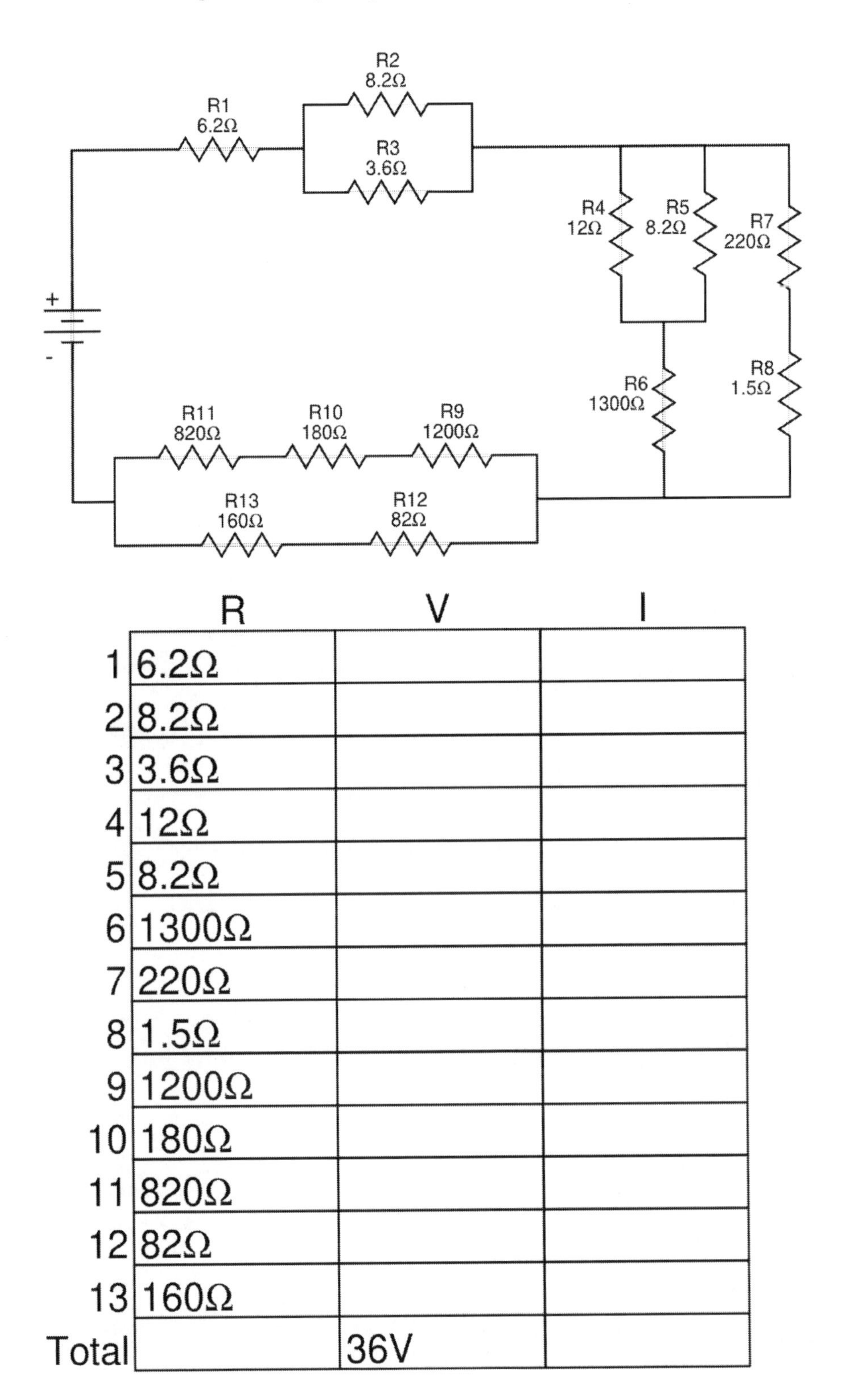

	R	V	I
1	6.2Ω		
2	8.2Ω		
3	3.6Ω		
4	12Ω		
5	8.2Ω		
6	1300Ω		
7	220Ω		
8	1.5Ω		
9	1200Ω		
10	180Ω		
11	820Ω		
12	82Ω		
13	160Ω		
Total		36V	

Exercise #73

Find the voltage and amperage for each resistor.

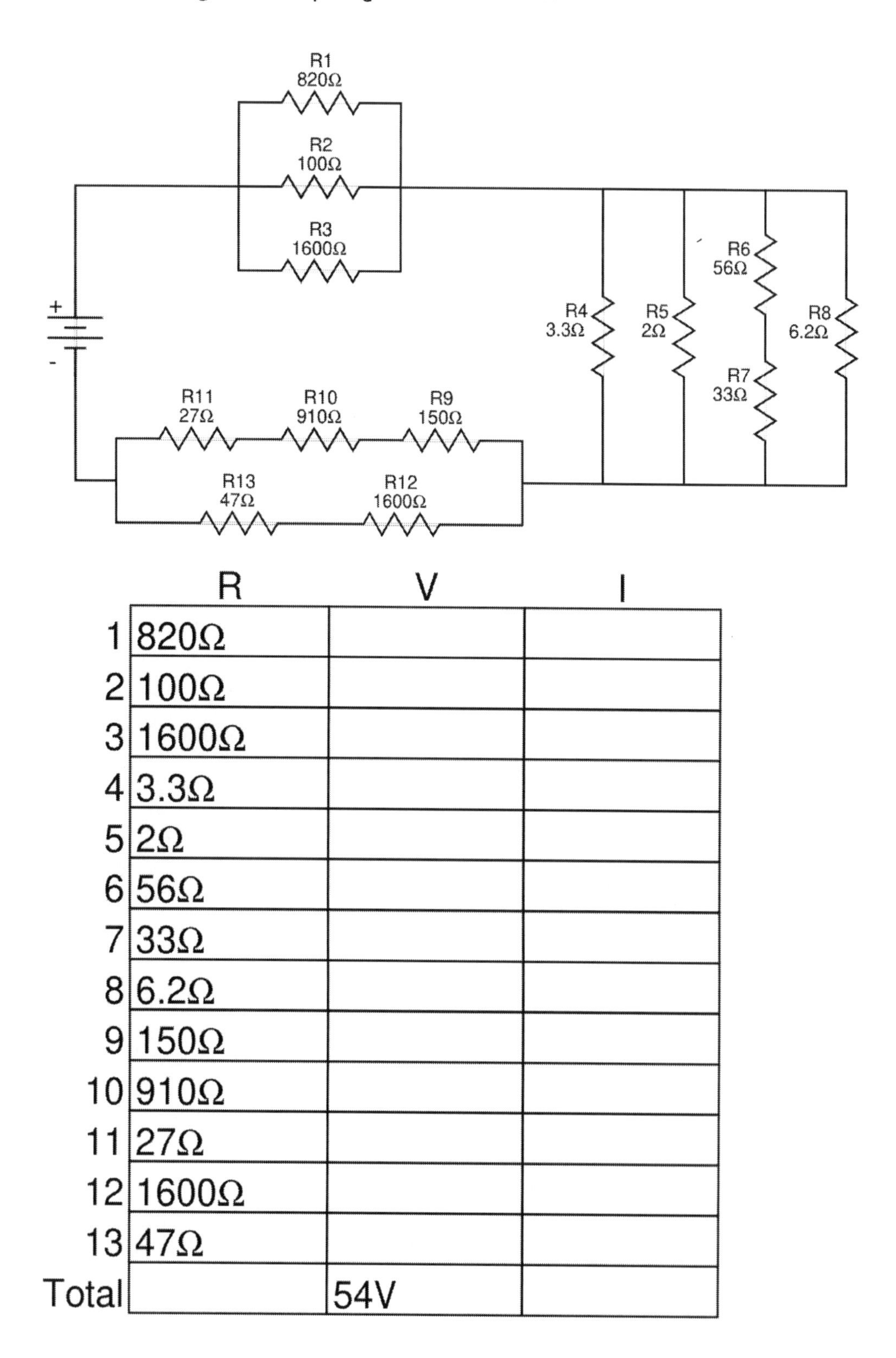

	R	V	I
1	820Ω		
2	100Ω		
3	1600Ω		
4	3.3Ω		
5	2Ω		
6	56Ω		
7	33Ω		
8	6.2Ω		
9	150Ω		
10	910Ω		
11	27Ω		
12	1600Ω		
13	47Ω		
Total		54V	

Exercise #74

Find the voltage and amperage for each resistor.

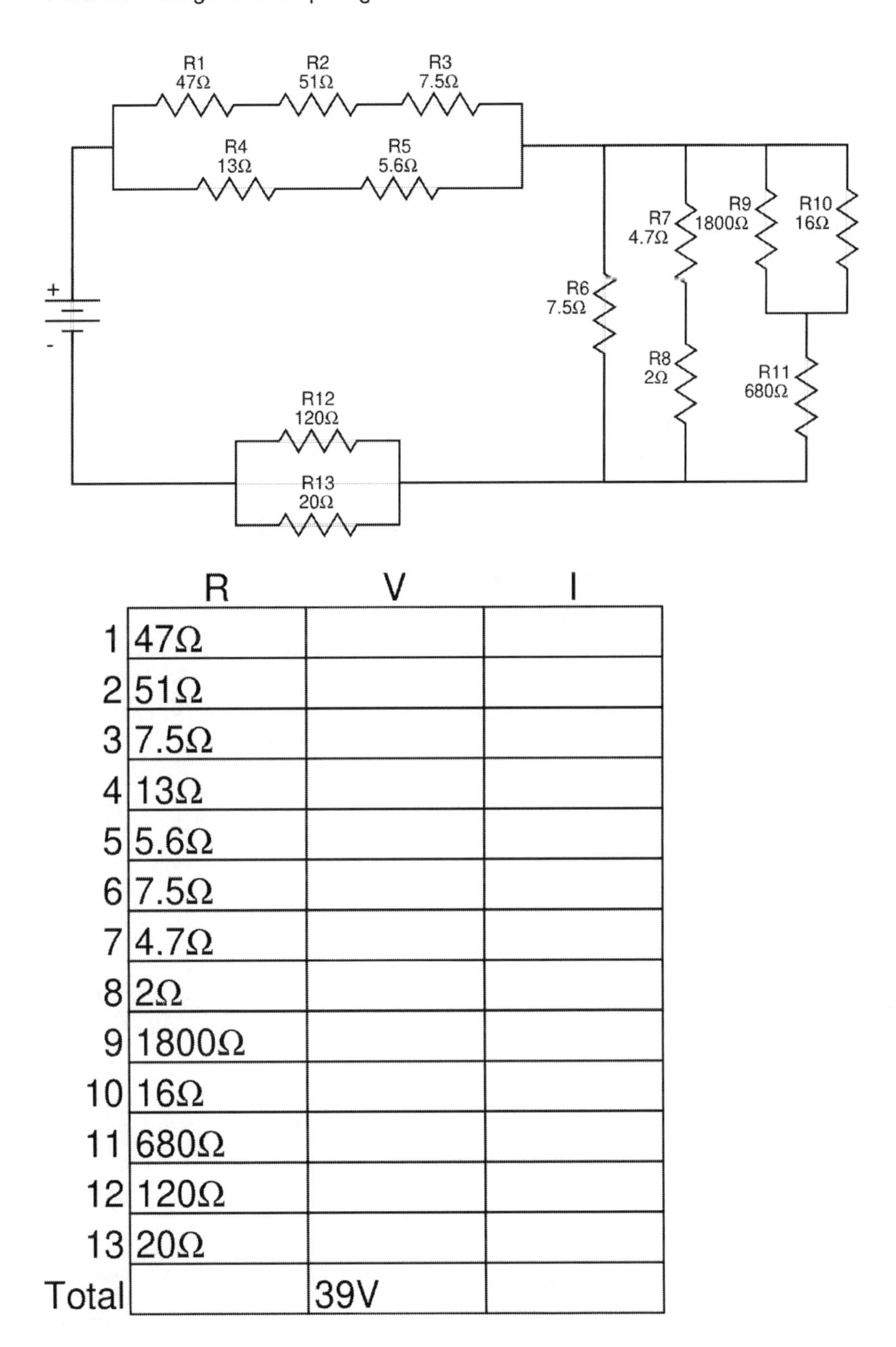

	R	V	I
1	47Ω		
2	51Ω		
3	7.5Ω		
4	13Ω		
5	5.6Ω		
6	7.5Ω		
7	4.7Ω		
8	2Ω		
9	1800Ω		
10	16Ω		
11	680Ω		
12	120Ω		
13	20Ω		
Total		39V	

Exercise #75

Find the voltage and amperage for each resistor.

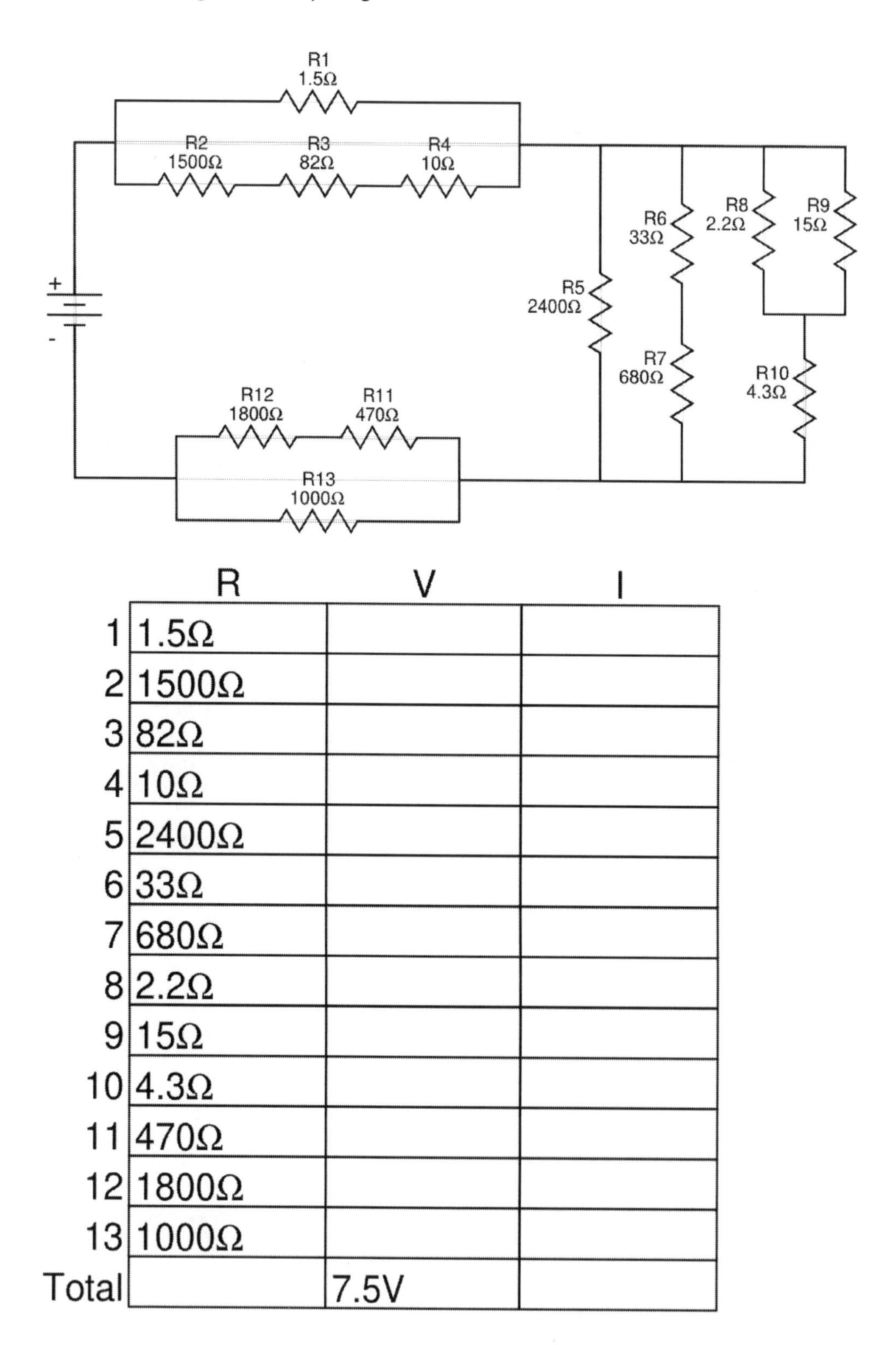

	R	V	I
1	1.5Ω		
2	1500Ω		
3	82Ω		
4	10Ω		
5	2400Ω		
6	33Ω		
7	680Ω		
8	2.2Ω		
9	15Ω		
10	4.3Ω		
11	470Ω		
12	1800Ω		
13	1000Ω		
Total		7.5V	

Exercise #76

Find the voltage and amperage for each resistor.

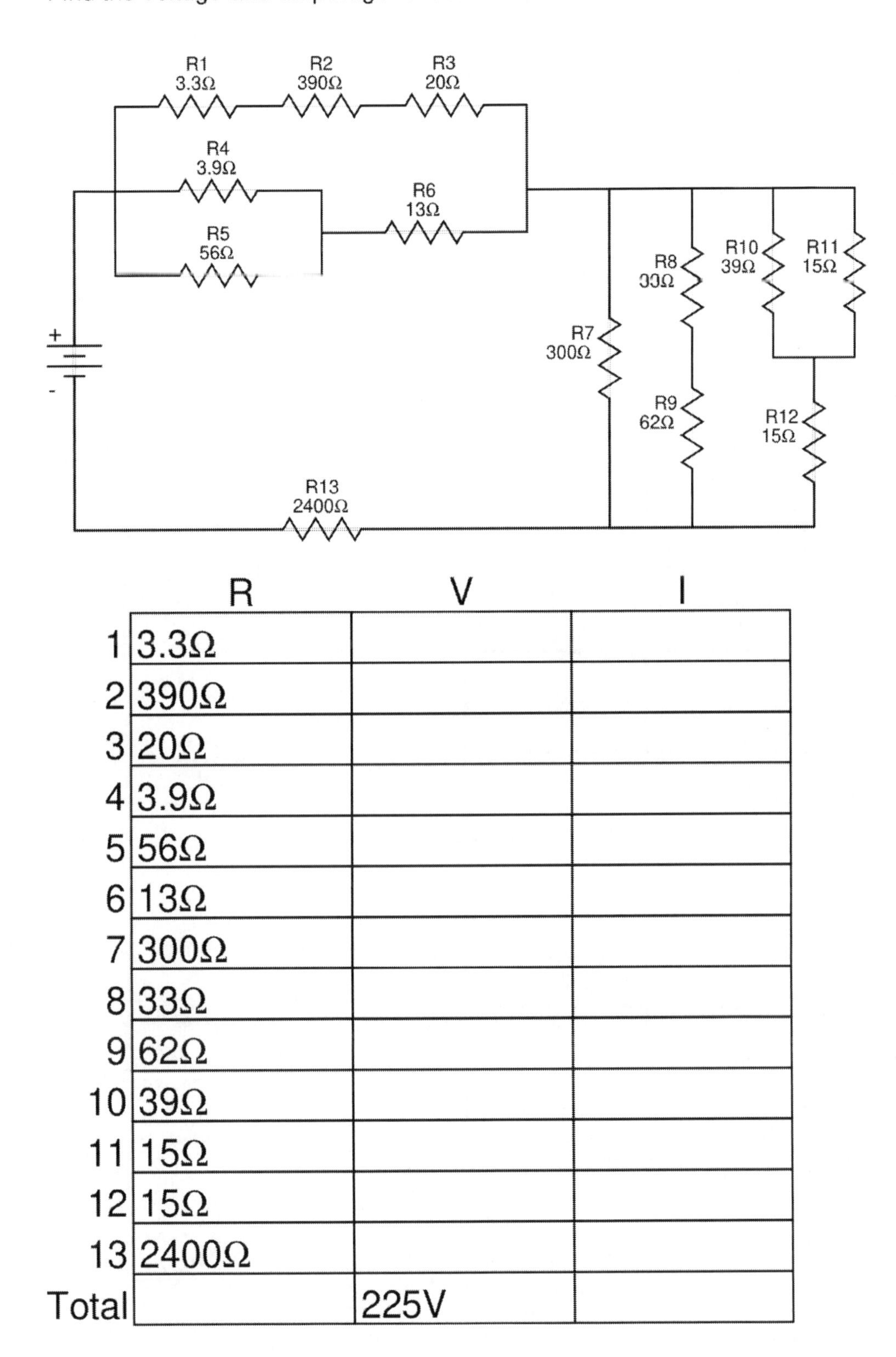

	R	V	I
1	3.3Ω		
2	390Ω		
3	20Ω		
4	3.9Ω		
5	56Ω		
6	13Ω		
7	300Ω		
8	33Ω		
9	62Ω		
10	39Ω		
11	15Ω		
12	15Ω		
13	2400Ω		
Total		225V	

Exercise #77

Find the voltage and amperage for each resistor.

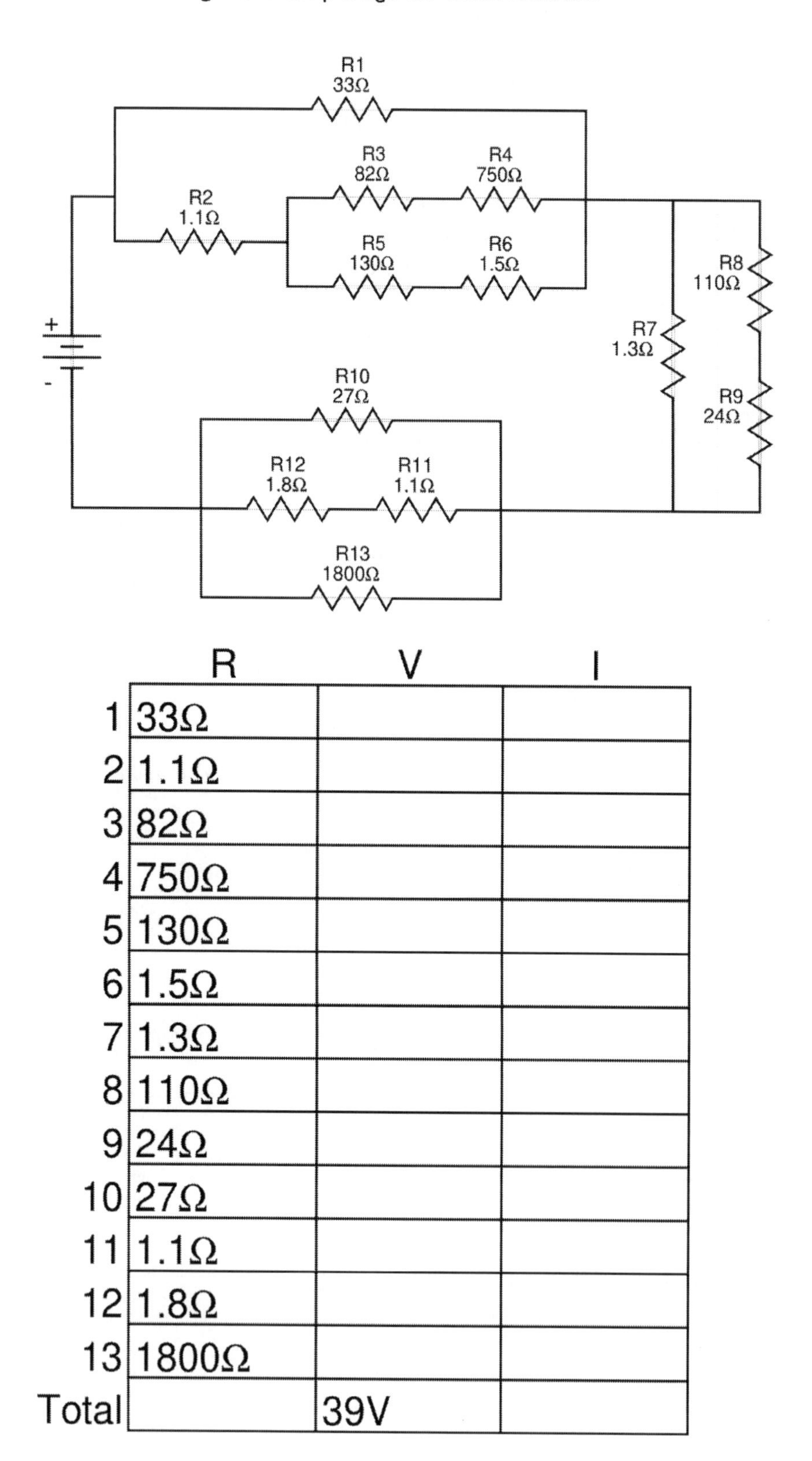

	R	V	I
1	33Ω		
2	1.1Ω		
3	82Ω		
4	750Ω		
5	130Ω		
6	1.5Ω		
7	1.3Ω		
8	110Ω		
9	24Ω		
10	27Ω		
11	1.1Ω		
12	1.8Ω		
13	1800Ω		
Total		39V	

Exercise #78

Find the voltage and amperage for each resistor.

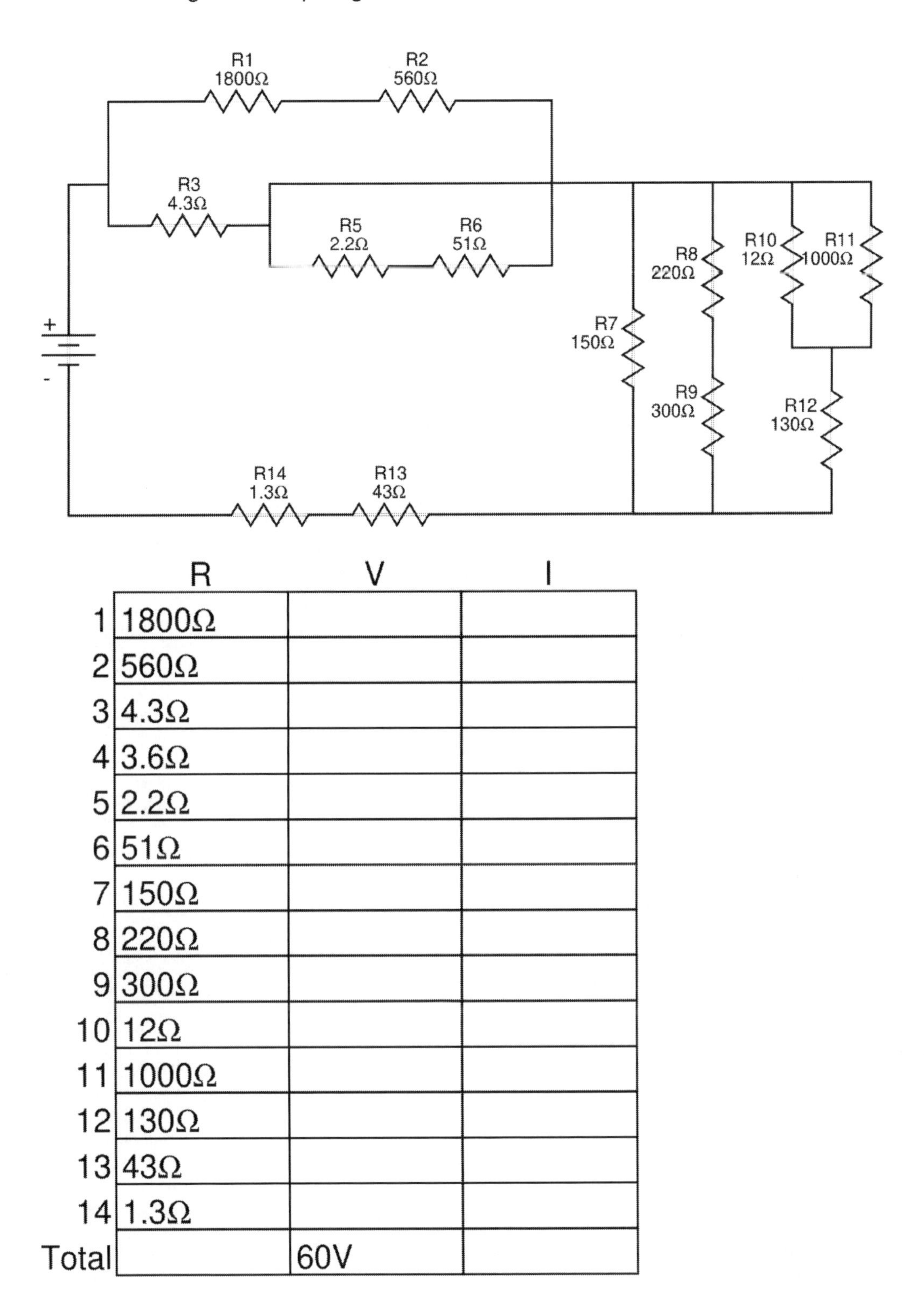

	R	V	I
1	1800Ω		
2	560Ω		
3	4.3Ω		
4	3.6Ω		
5	2.2Ω		
6	51Ω		
7	150Ω		
8	220Ω		
9	300Ω		
10	12Ω		
11	1000Ω		
12	130Ω		
13	43Ω		
14	1.3Ω		
Total		60V	

Exercise #79

Find the voltage and amperage for each resistor.

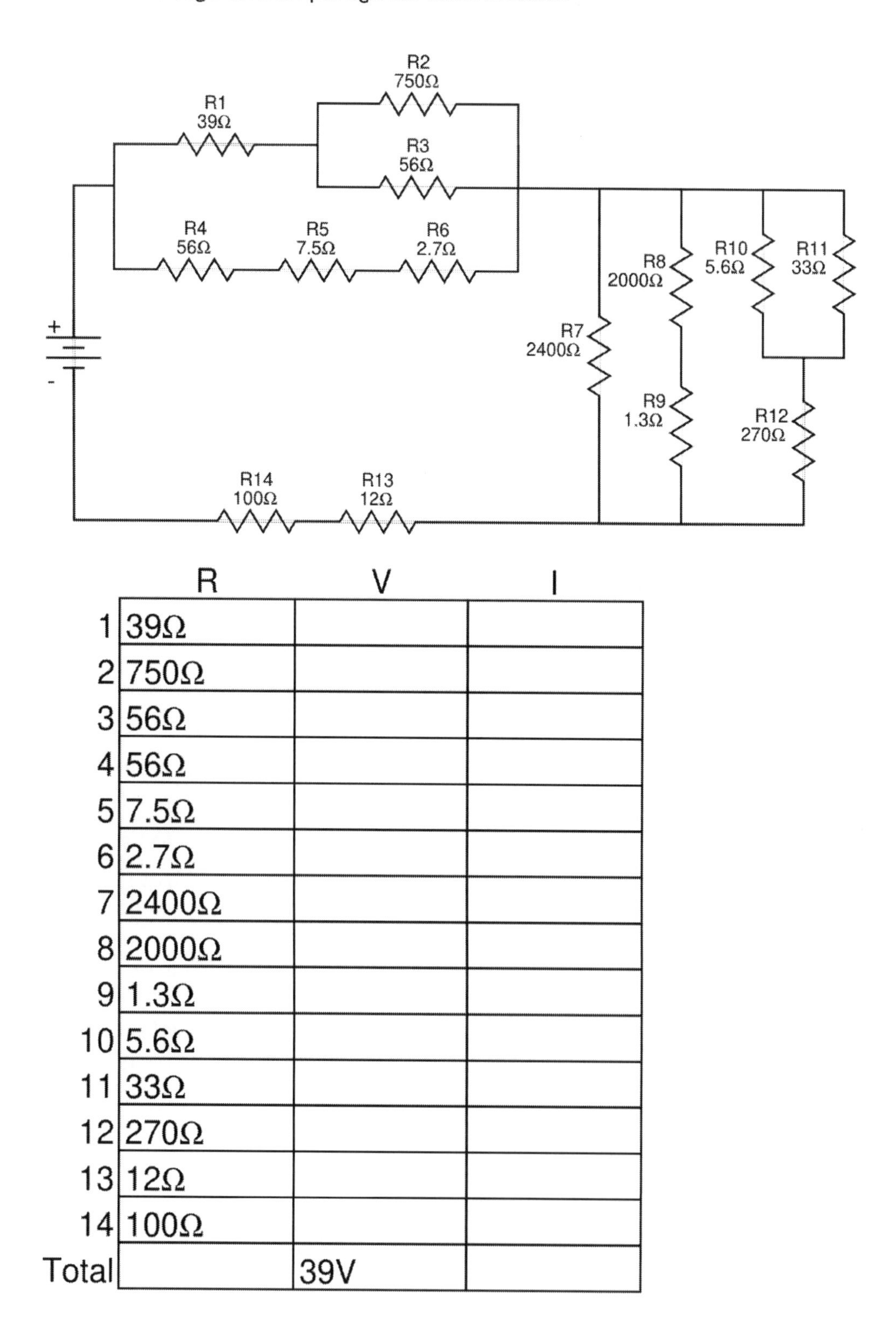

	R	V	I
1	39Ω		
2	750Ω		
3	56Ω		
4	56Ω		
5	7.5Ω		
6	2.7Ω		
7	2400Ω		
8	2000Ω		
9	1.3Ω		
10	5.6Ω		
11	33Ω		
12	270Ω		
13	12Ω		
14	100Ω		
Total		39V	

Exercise #80

Find the voltage and amperage for each resistor.

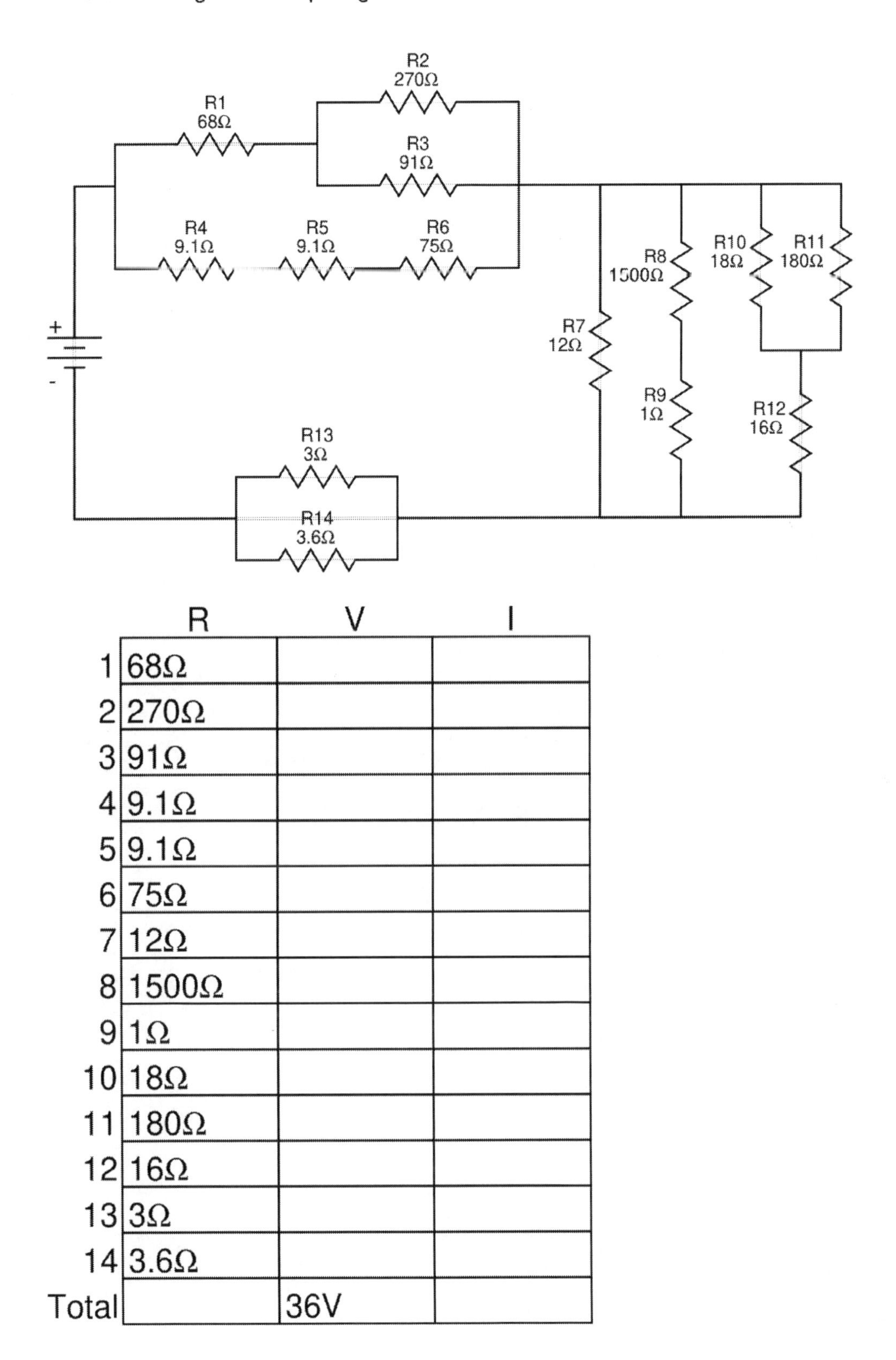

	R	V	I
1	68Ω		
2	270Ω		
3	91Ω		
4	9.1Ω		
5	9.1Ω		
6	75Ω		
7	12Ω		
8	1500Ω		
9	1Ω		
10	18Ω		
11	180Ω		
12	16Ω		
13	3Ω		
14	3.6Ω		
Total		36V	

Exercise #81

Find the voltage and amperage for each resistor.

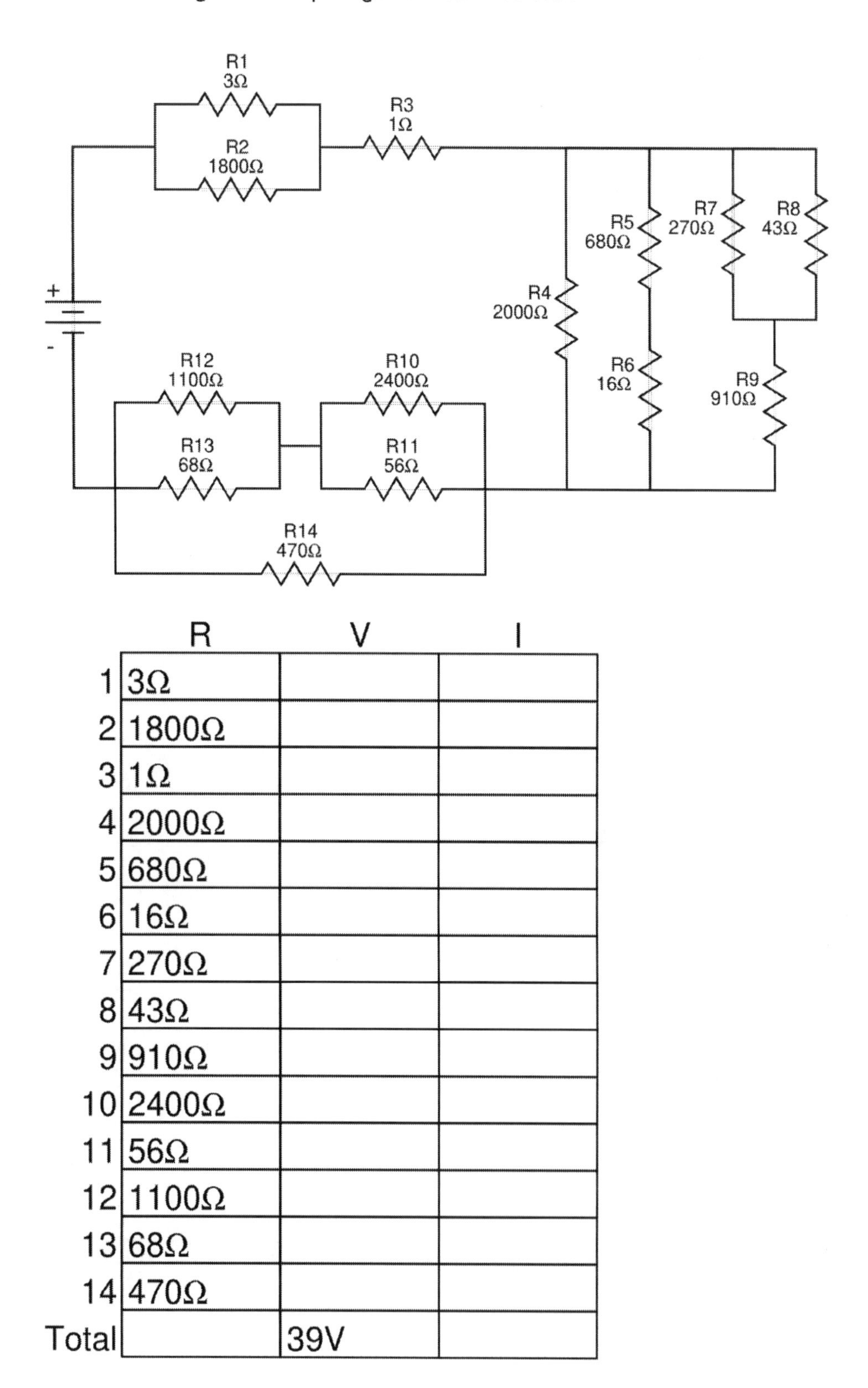

	R	V	I
1	3Ω		
2	1800Ω		
3	1Ω		
4	2000Ω		
5	680Ω		
6	16Ω		
7	270Ω		
8	43Ω		
9	910Ω		
10	2400Ω		
11	56Ω		
12	1100Ω		
13	68Ω		
14	470Ω		
Total		39V	

Exercise #82

Find the voltage and amperage for each resistor.

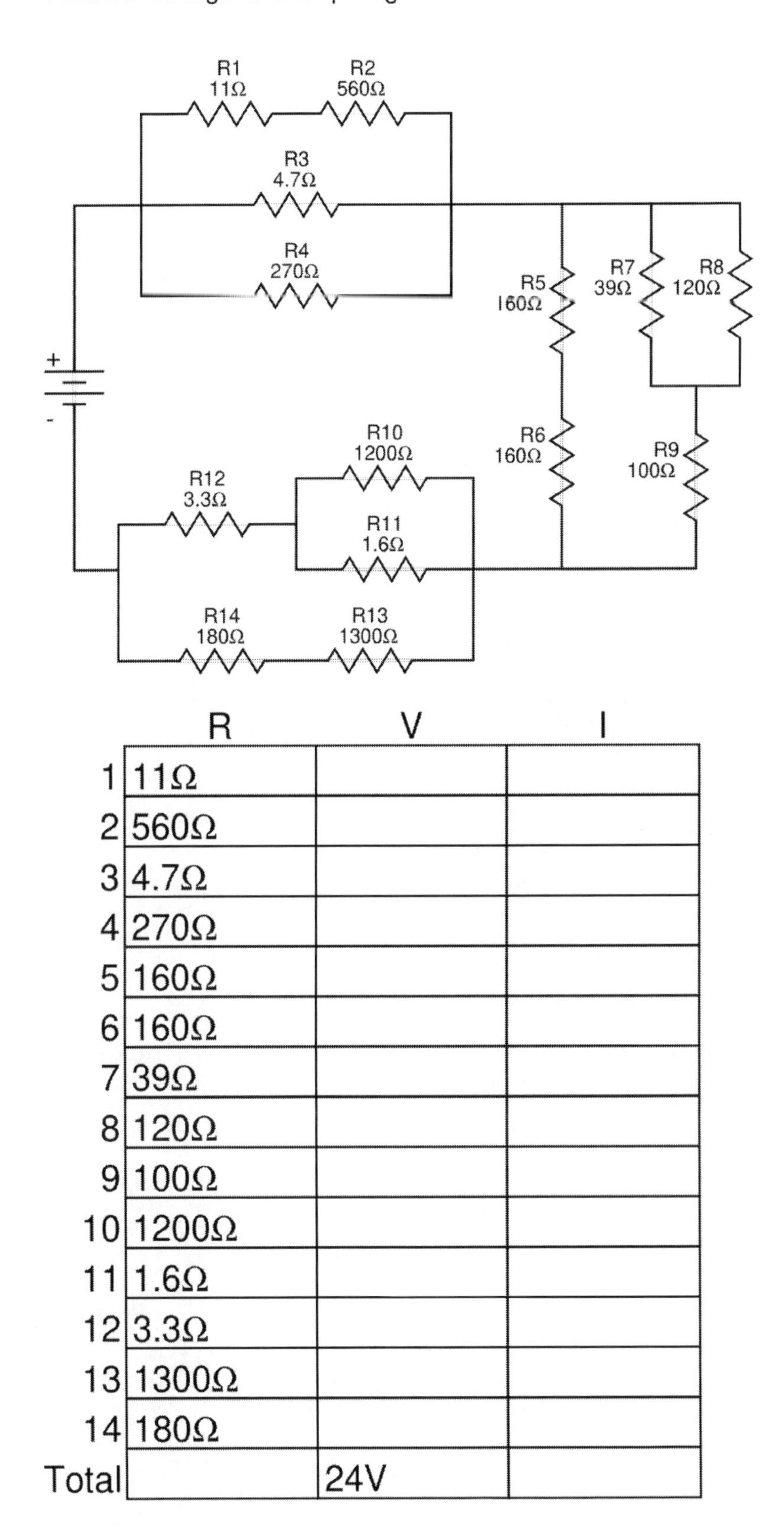

	R	V	I
1	11Ω		
2	560Ω		
3	4.7Ω		
4	270Ω		
5	160Ω		
6	160Ω		
7	39Ω		
8	120Ω		
9	100Ω		
10	1200Ω		
11	1.6Ω		
12	3.3Ω		
13	1300Ω		
14	180Ω		
Total		24V	

Exercise #83

Find the voltage and amperage for each resistor.

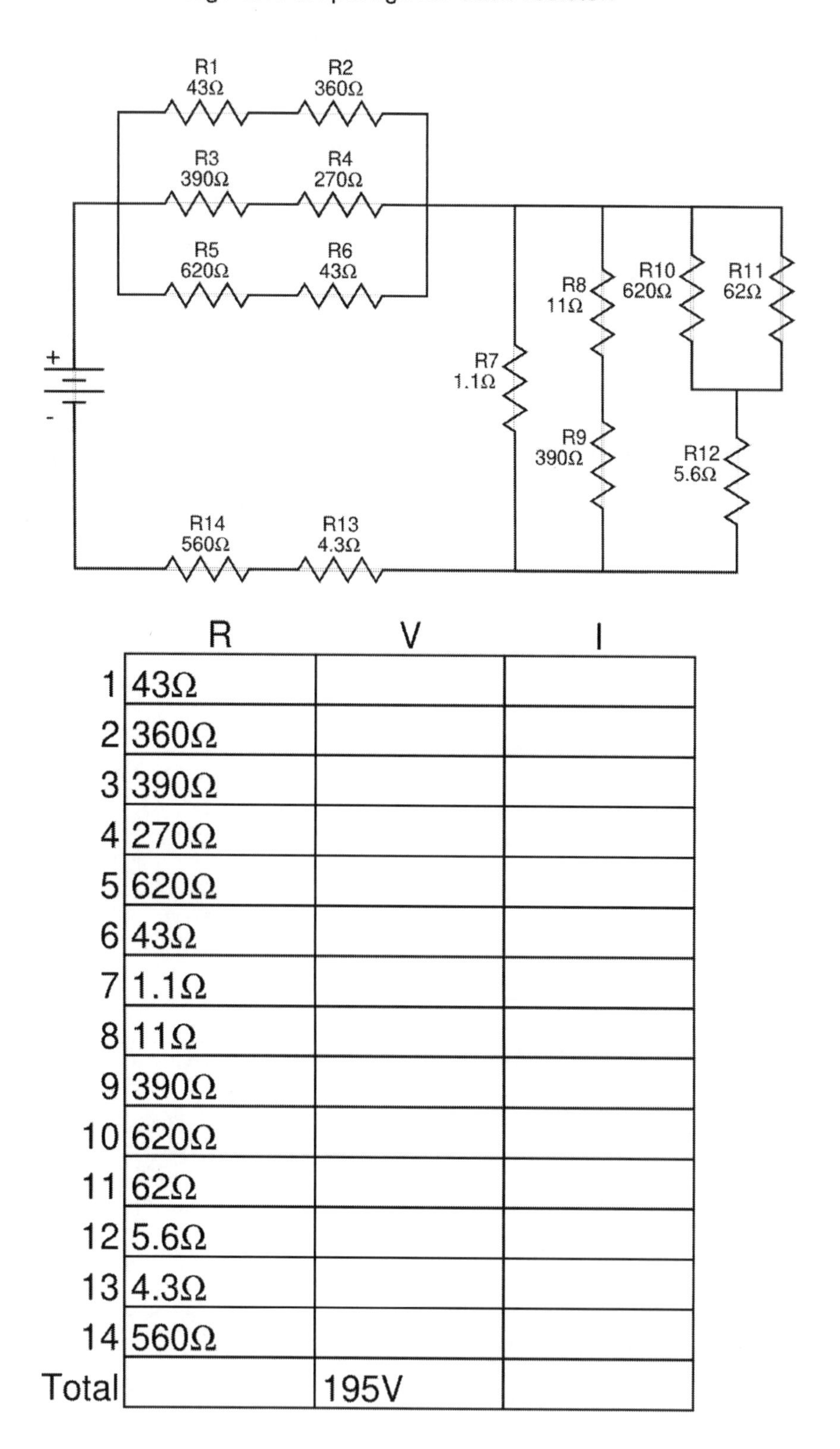

	R	V	I
1	43Ω		
2	360Ω		
3	390Ω		
4	270Ω		
5	620Ω		
6	43Ω		
7	1.1Ω		
8	11Ω		
9	390Ω		
10	620Ω		
11	62Ω		
12	5.6Ω		
13	4.3Ω		
14	560Ω		
Total		195V	

Exercise #84

Find the voltage and amperage for each resistor.

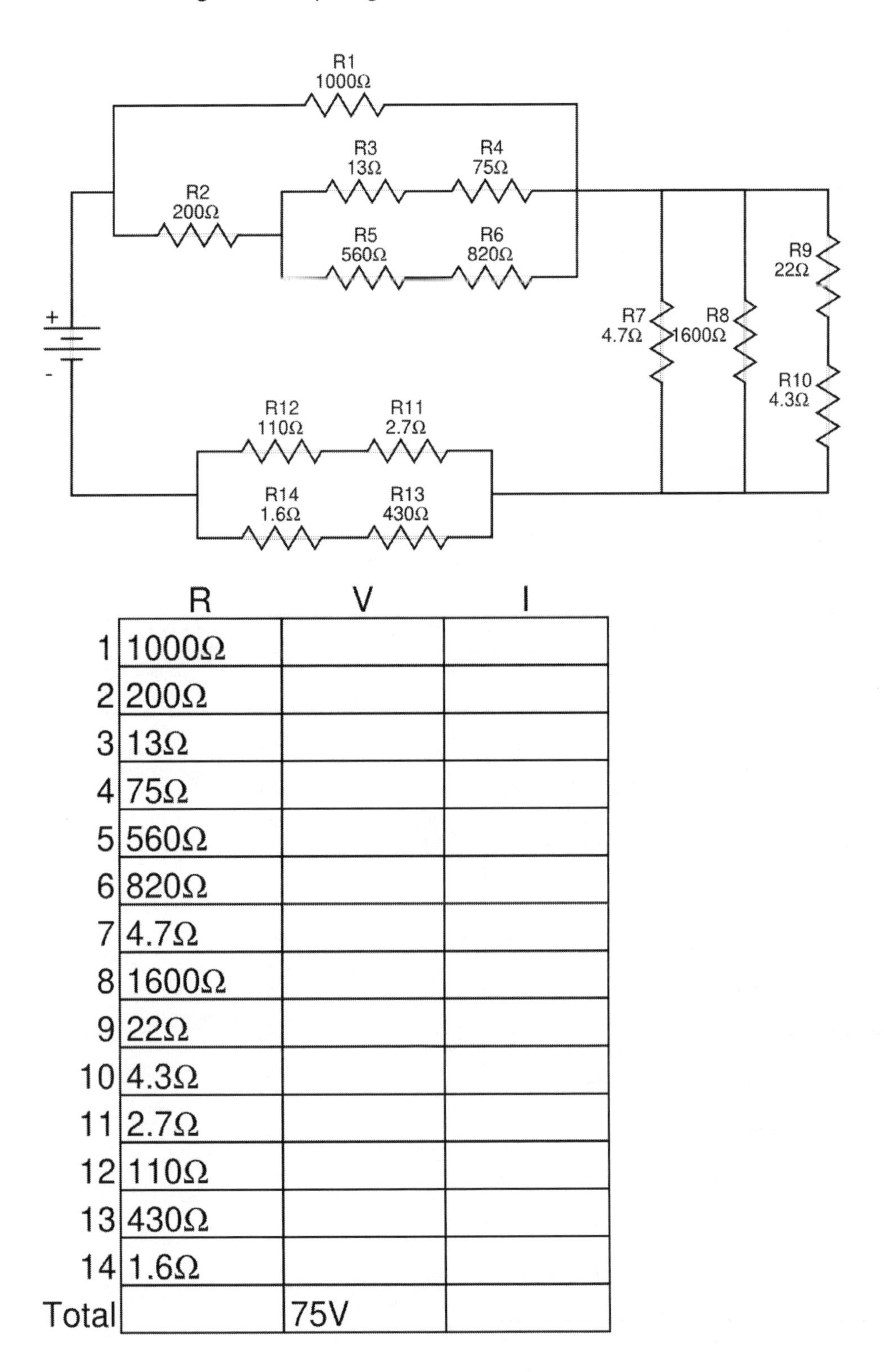

	R	V	I
1	1000Ω		
2	200Ω		
3	13Ω		
4	75Ω		
5	560Ω		
6	820Ω		
7	4.7Ω		
8	1600Ω		
9	22Ω		
10	4.3Ω		
11	2.7Ω		
12	110Ω		
13	430Ω		
14	1.6Ω		
Total		75V	

Exercise #85

Find the voltage and amperage for each resistor.

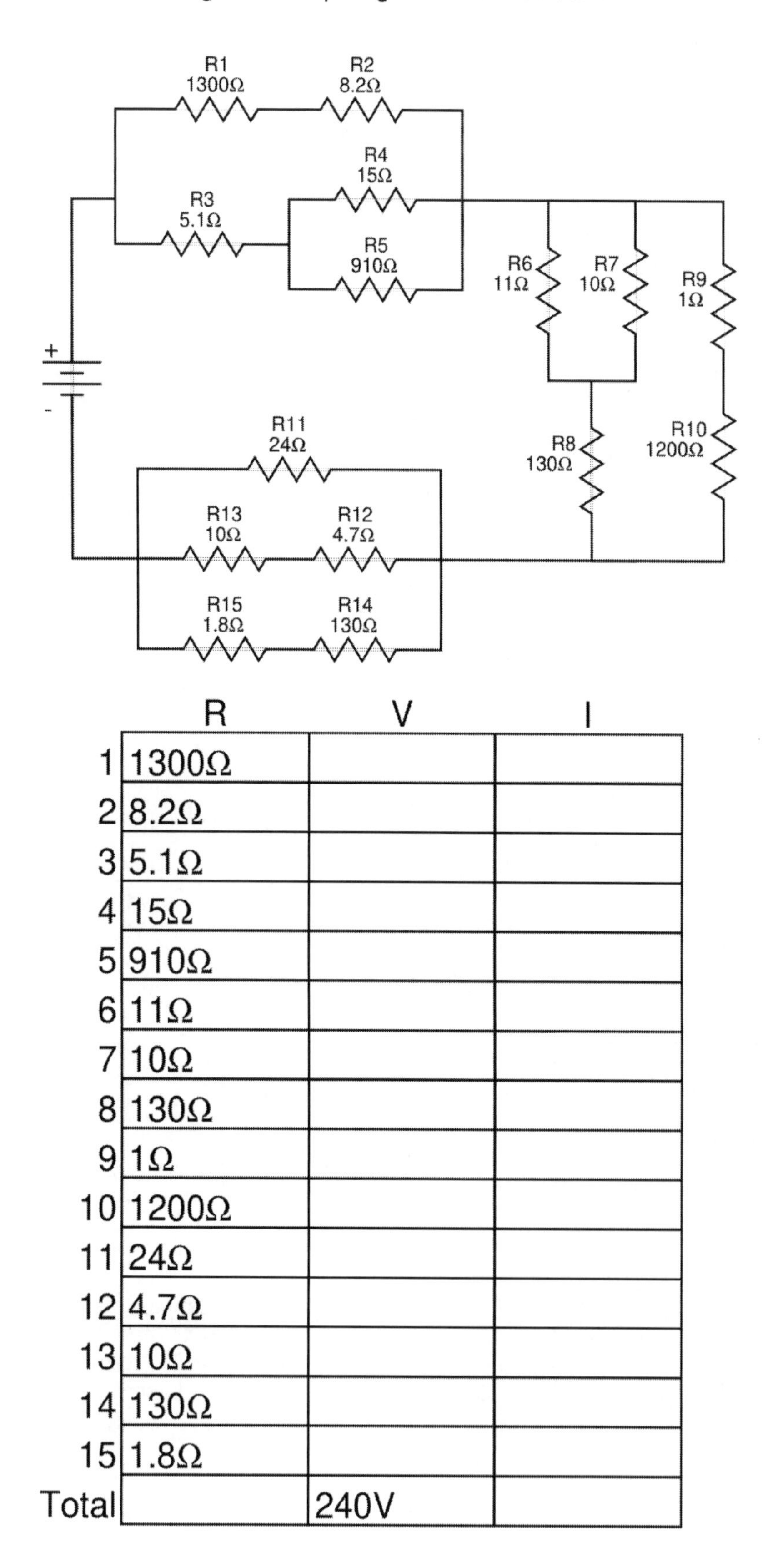

	R	V	I
1	1300Ω		
2	8.2Ω		
3	5.1Ω		
4	15Ω		
5	910Ω		
6	11Ω		
7	10Ω		
8	130Ω		
9	1Ω		
10	1200Ω		
11	24Ω		
12	4.7Ω		
13	10Ω		
14	130Ω		
15	1.8Ω		
Total		240V	

Exercise #86

Find the voltage and amperage for each resistor.

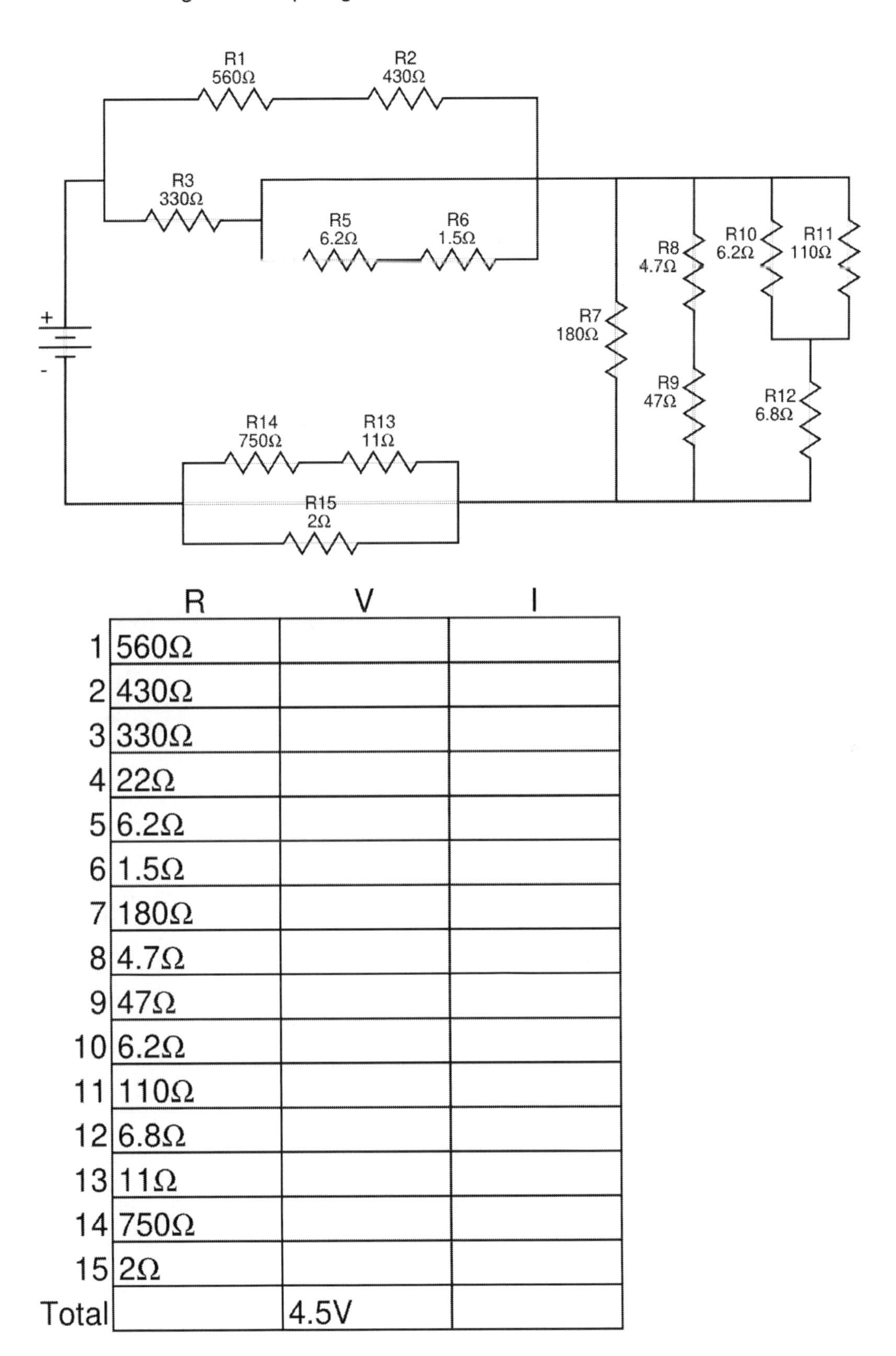

	R	V	I
1	560Ω		
2	430Ω		
3	330Ω		
4	22Ω		
5	6.2Ω		
6	1.5Ω		
7	180Ω		
8	4.7Ω		
9	47Ω		
10	6.2Ω		
11	110Ω		
12	6.8Ω		
13	11Ω		
14	750Ω		
15	2Ω		
Total		4.5V	

Exercise #87

Find the voltage and amperage for each resistor.

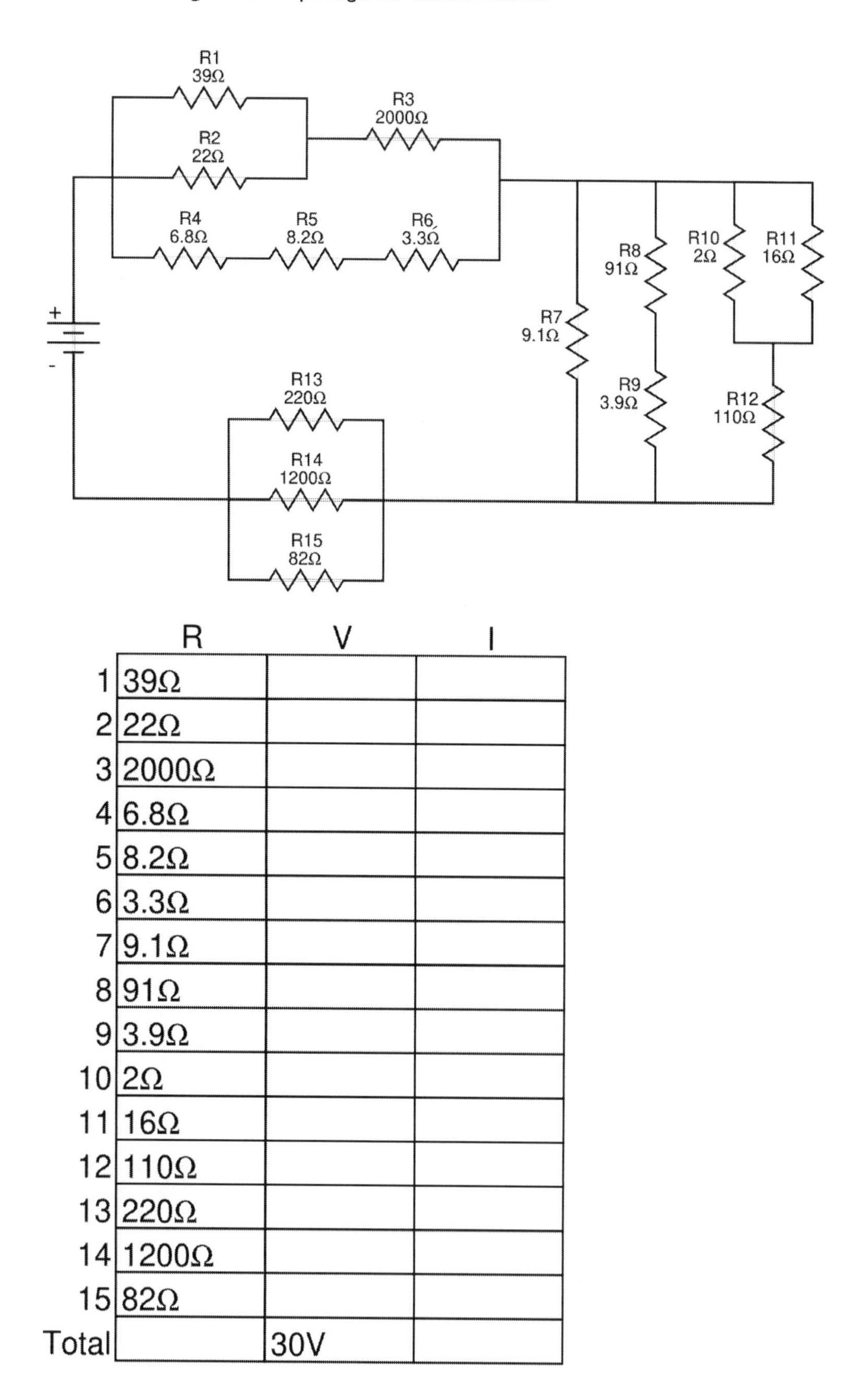

	R	V	I
1	39Ω		
2	22Ω		
3	2000Ω		
4	6.8Ω		
5	8.2Ω		
6	3.3Ω		
7	9.1Ω		
8	91Ω		
9	3.9Ω		
10	2Ω		
11	16Ω		
12	110Ω		
13	220Ω		
14	1200Ω		
15	82Ω		
Total		30V	

Exercise #88

Find the voltage and amperage for each resistor.

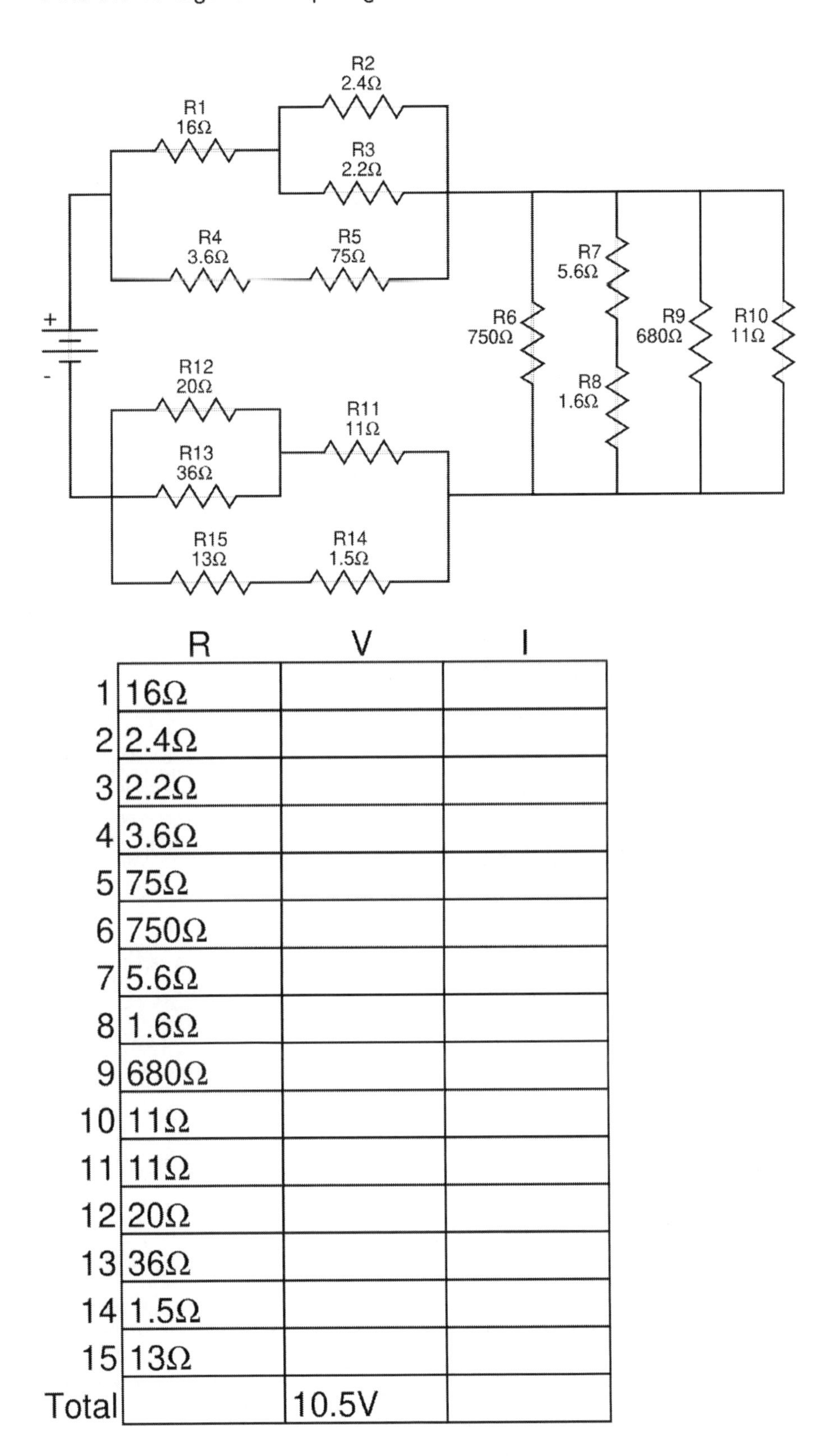

	R	V	I
1	16Ω		
2	2.4Ω		
3	2.2Ω		
4	3.6Ω		
5	75Ω		
6	750Ω		
7	5.6Ω		
8	1.6Ω		
9	680Ω		
10	11Ω		
11	11Ω		
12	20Ω		
13	36Ω		
14	1.5Ω		
15	13Ω		
Total		10.5V	

Exercise #89

Find the voltage and amperage for each resistor.

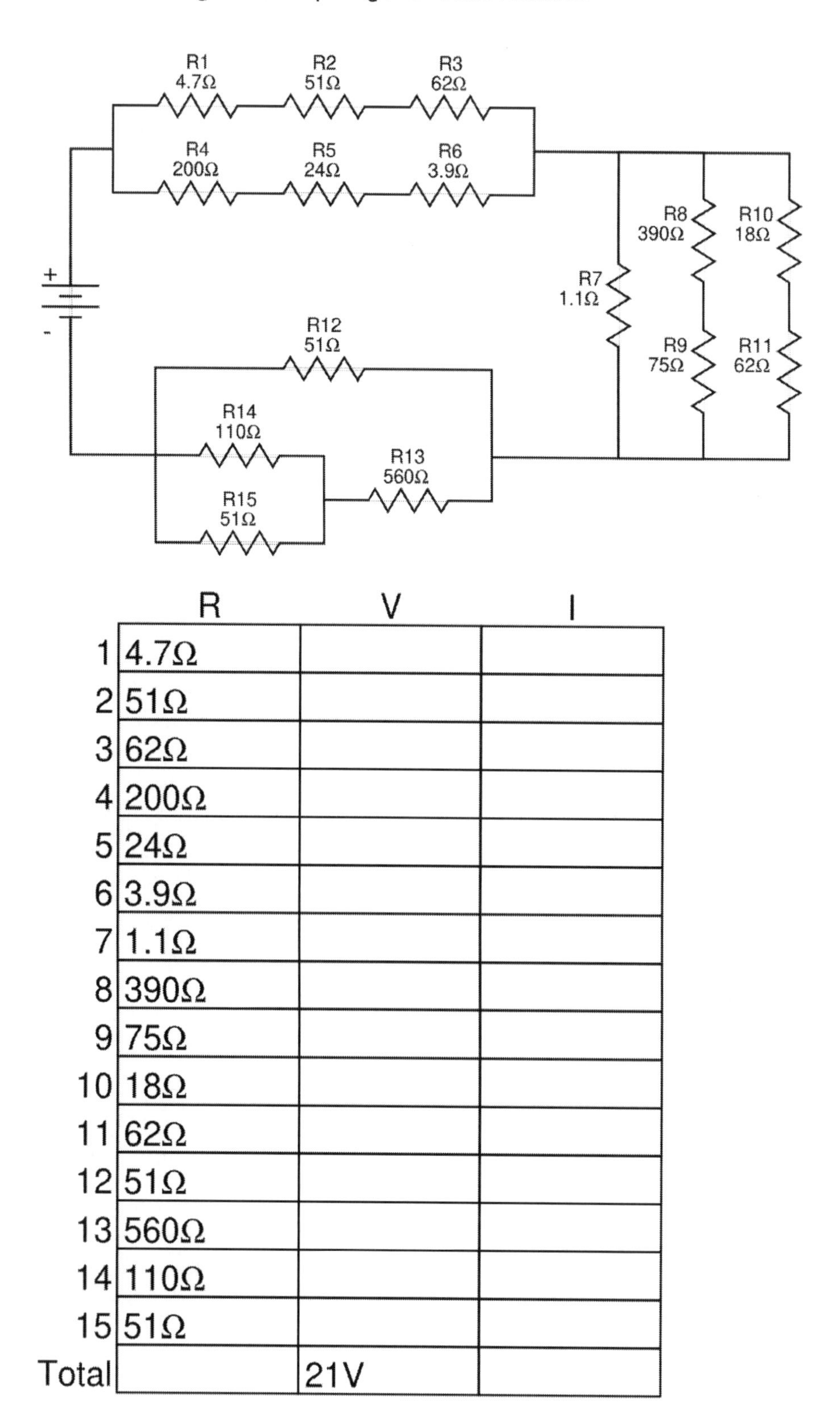

	R	V	I
1	4.7Ω		
2	51Ω		
3	62Ω		
4	200Ω		
5	24Ω		
6	3.9Ω		
7	1.1Ω		
8	390Ω		
9	75Ω		
10	18Ω		
11	62Ω		
12	51Ω		
13	560Ω		
14	110Ω		
15	51Ω		
Total		21V	

Exercise #90

Find the voltage and amperage for each resistor.

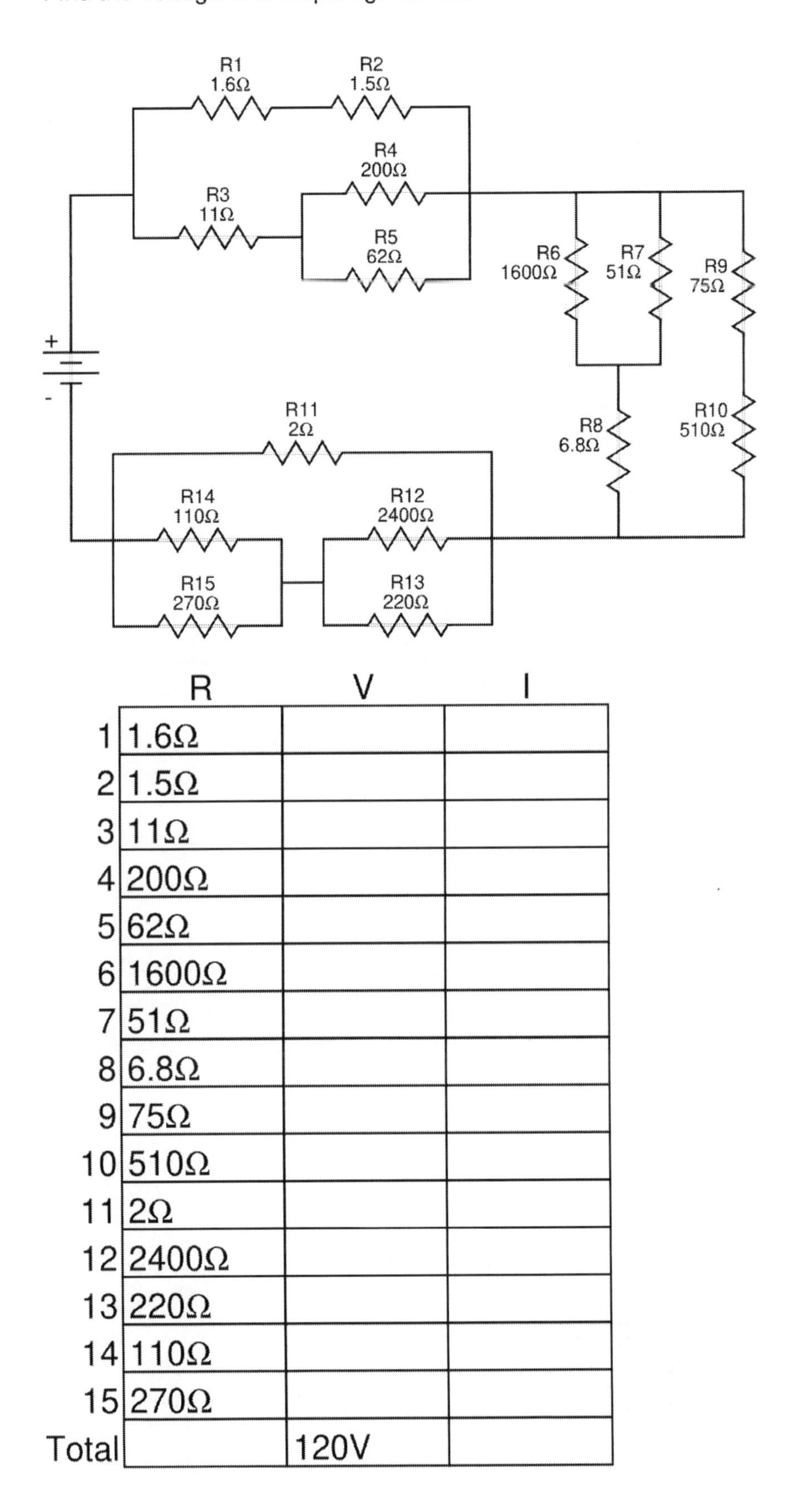

	R	V	I
1	1.6Ω		
2	1.5Ω		
3	11Ω		
4	200Ω		
5	62Ω		
6	1600Ω		
7	51Ω		
8	6.8Ω		
9	75Ω		
10	510Ω		
11	2Ω		
12	2400Ω		
13	220Ω		
14	110Ω		
15	270Ω		
Total		120V	

Exercise #91

Find the voltage and amperage for each resistor.

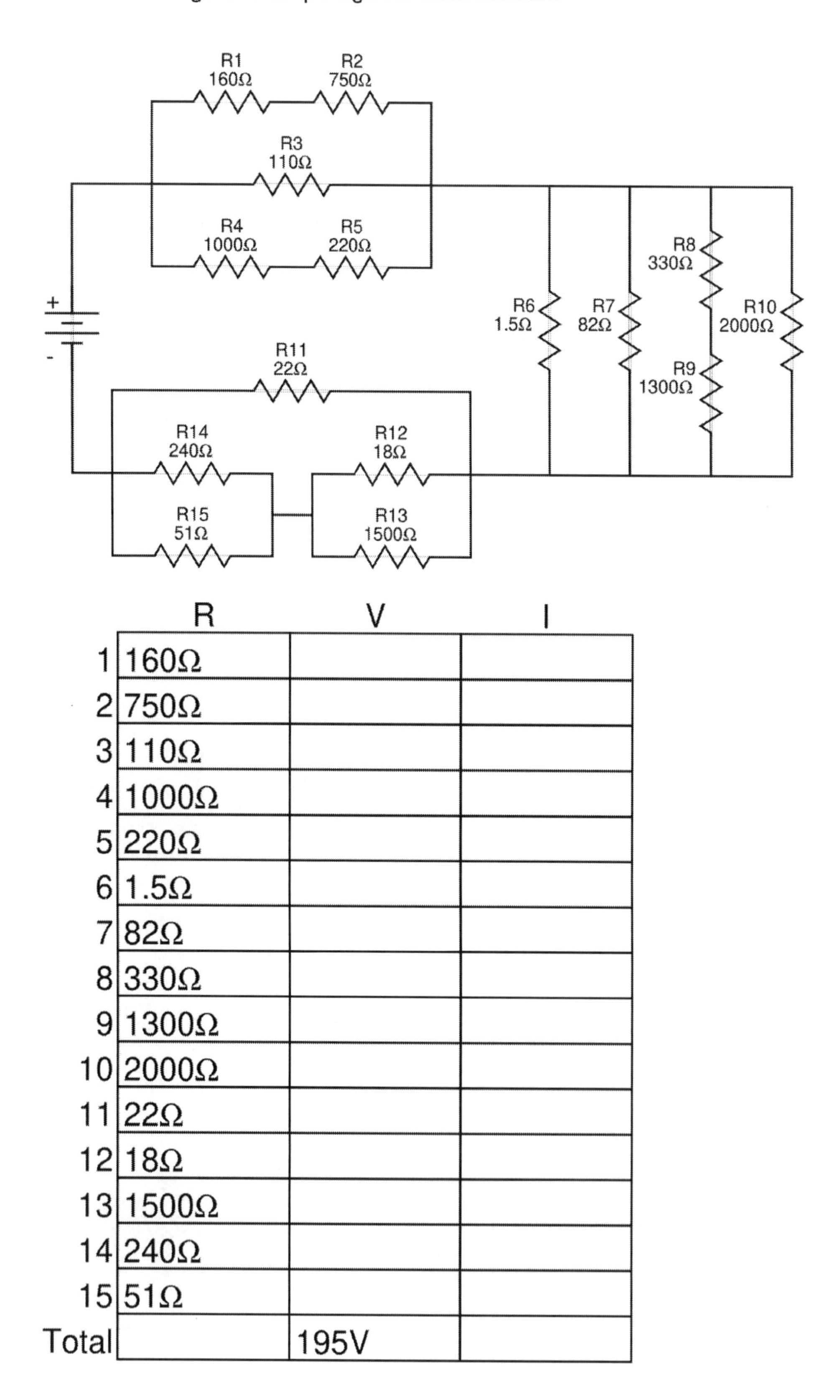

	R	V	I
1	160Ω		
2	750Ω		
3	110Ω		
4	1000Ω		
5	220Ω		
6	1.5Ω		
7	82Ω		
8	330Ω		
9	1300Ω		
10	2000Ω		
11	22Ω		
12	18Ω		
13	1500Ω		
14	240Ω		
15	51Ω		
Total		195V	

Exercise #92

Find the voltage and amperage for each resistor.

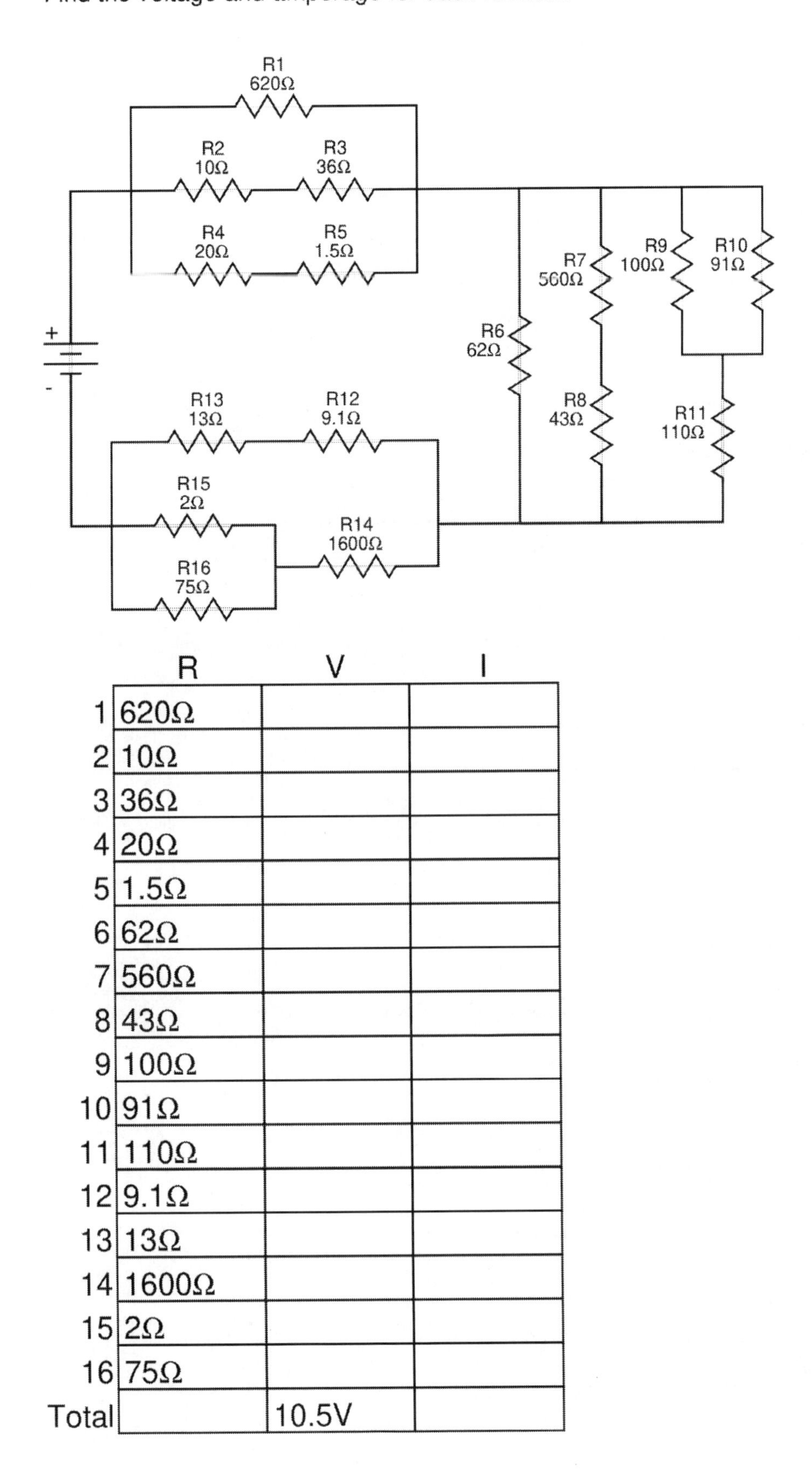

	R	V	I
1	620Ω		
2	10Ω		
3	36Ω		
4	20Ω		
5	1.5Ω		
6	62Ω		
7	560Ω		
8	43Ω		
9	100Ω		
10	91Ω		
11	110Ω		
12	9.1Ω		
13	13Ω		
14	1600Ω		
15	2Ω		
16	75Ω		
Total		10.5V	

Exercise #93

Find the voltage and amperage for each resistor.

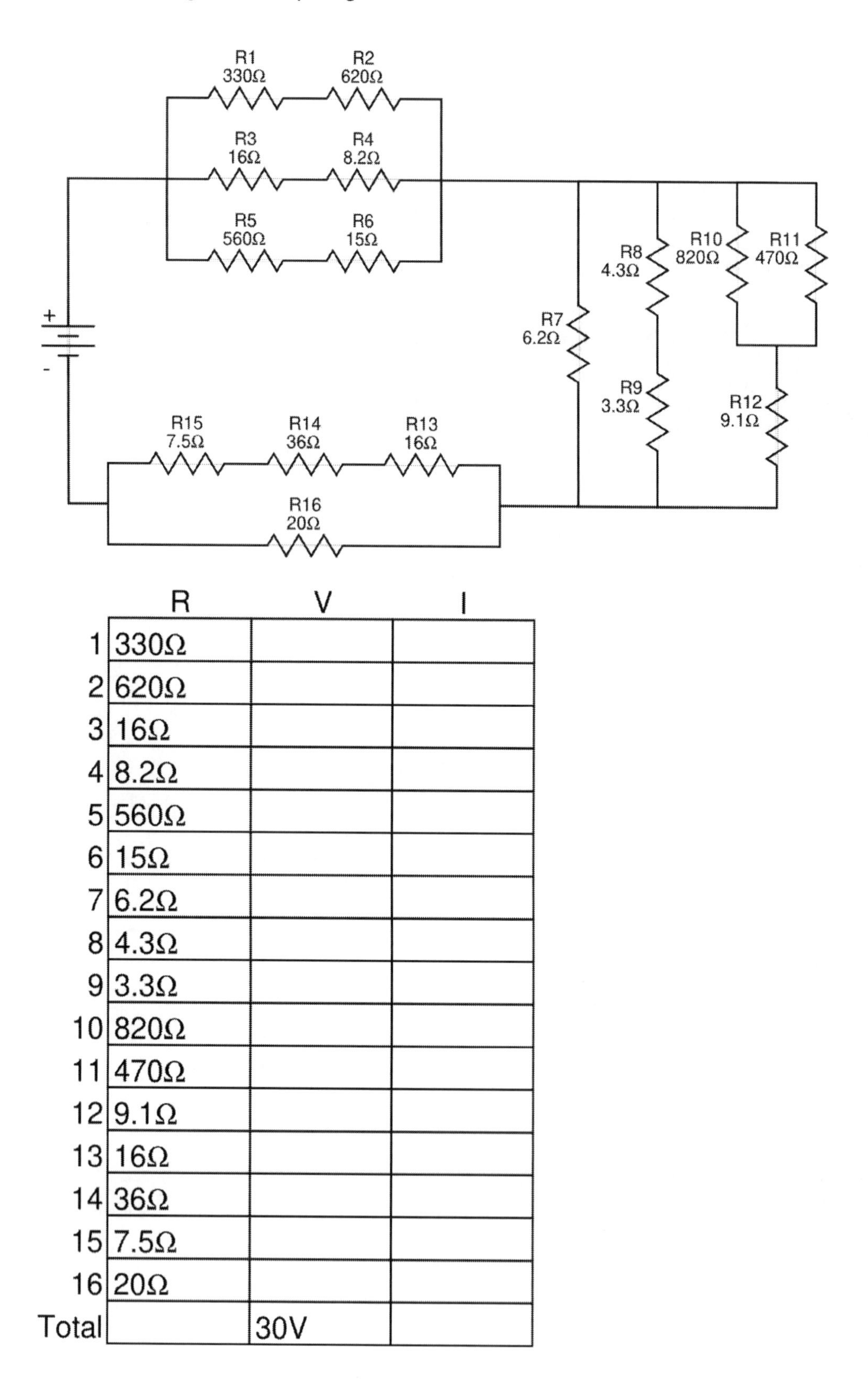

	R	V	I
1	330Ω		
2	620Ω		
3	16Ω		
4	8.2Ω		
5	560Ω		
6	15Ω		
7	6.2Ω		
8	4.3Ω		
9	3.3Ω		
10	820Ω		
11	470Ω		
12	9.1Ω		
13	16Ω		
14	36Ω		
15	7.5Ω		
16	20Ω		
Total		30V	

Exercise #94

Find the voltage and amperage for each resistor.

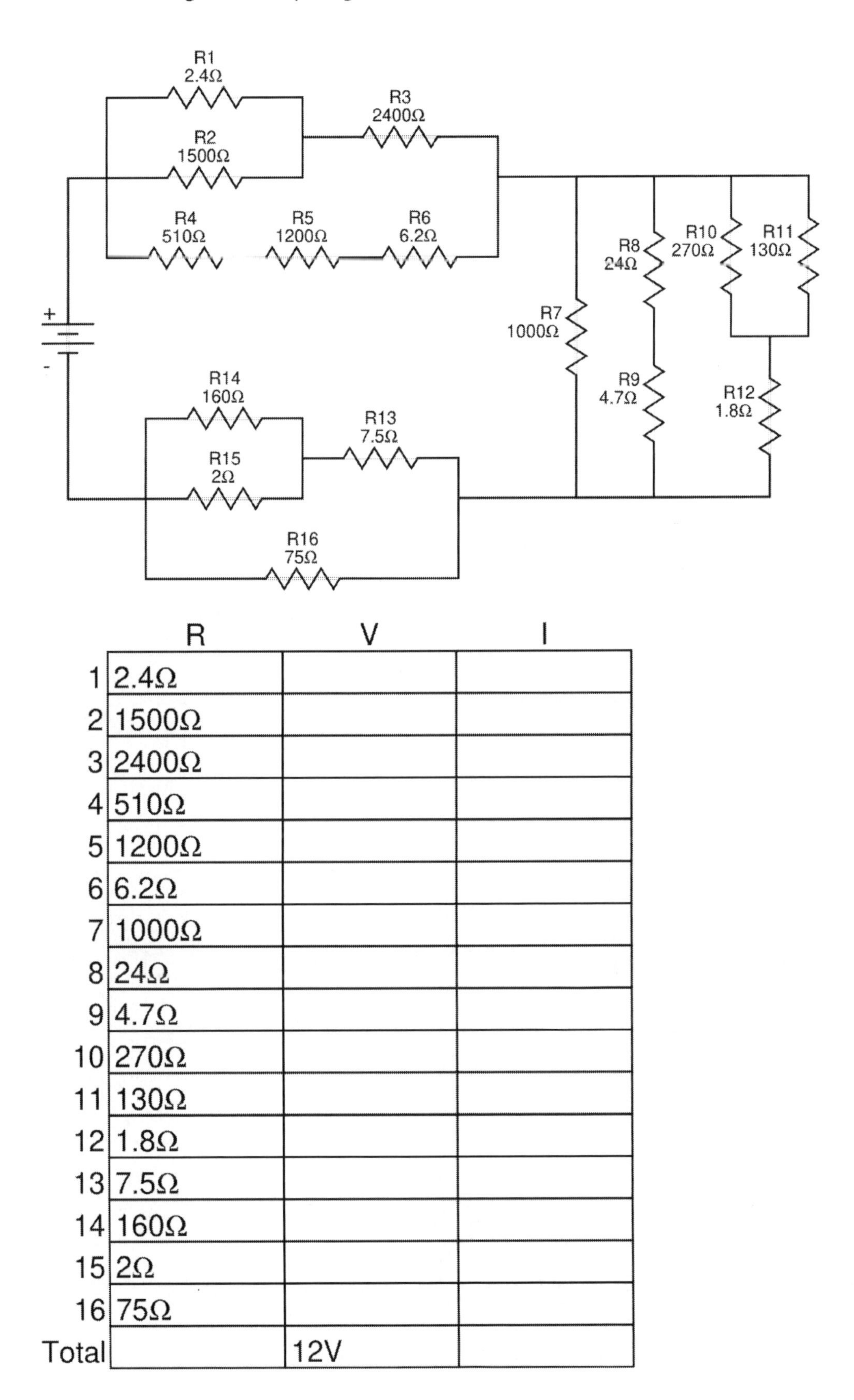

	R	V	I
1	2.4Ω		
2	1500Ω		
3	2400Ω		
4	510Ω		
5	1200Ω		
6	6.2Ω		
7	1000Ω		
8	24Ω		
9	4.7Ω		
10	270Ω		
11	130Ω		
12	1.8Ω		
13	7.5Ω		
14	160Ω		
15	2Ω		
16	75Ω		
Total		12V	

Exercise #95

Find the voltage and amperage for each resistor.

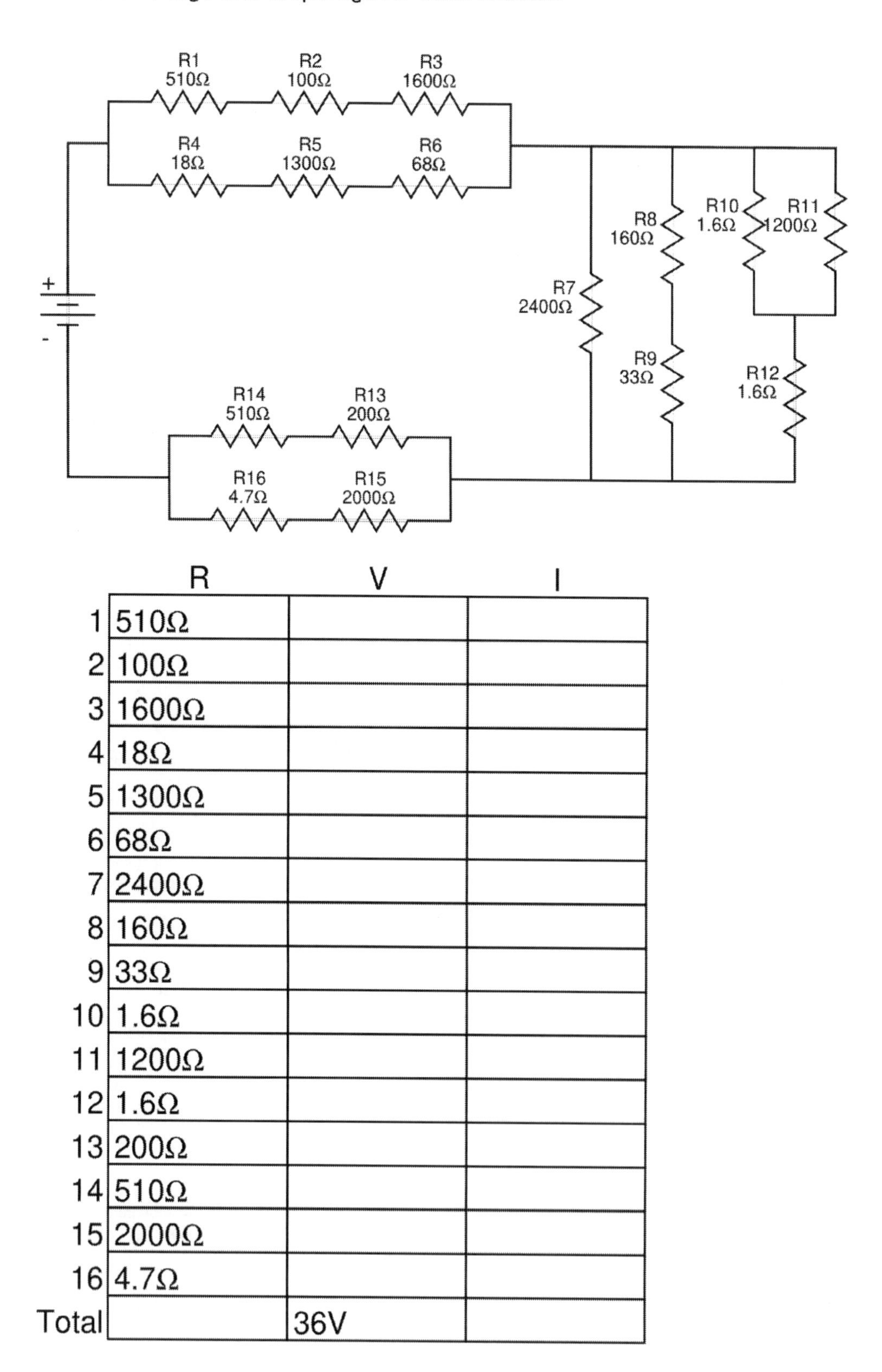

	R	V	I
1	510Ω		
2	100Ω		
3	1600Ω		
4	18Ω		
5	1300Ω		
6	68Ω		
7	2400Ω		
8	160Ω		
9	33Ω		
10	1.6Ω		
11	1200Ω		
12	1.6Ω		
13	200Ω		
14	510Ω		
15	2000Ω		
16	4.7Ω		
Total		36V	

Exercise #96

Find the voltage and amperage for each resistor.

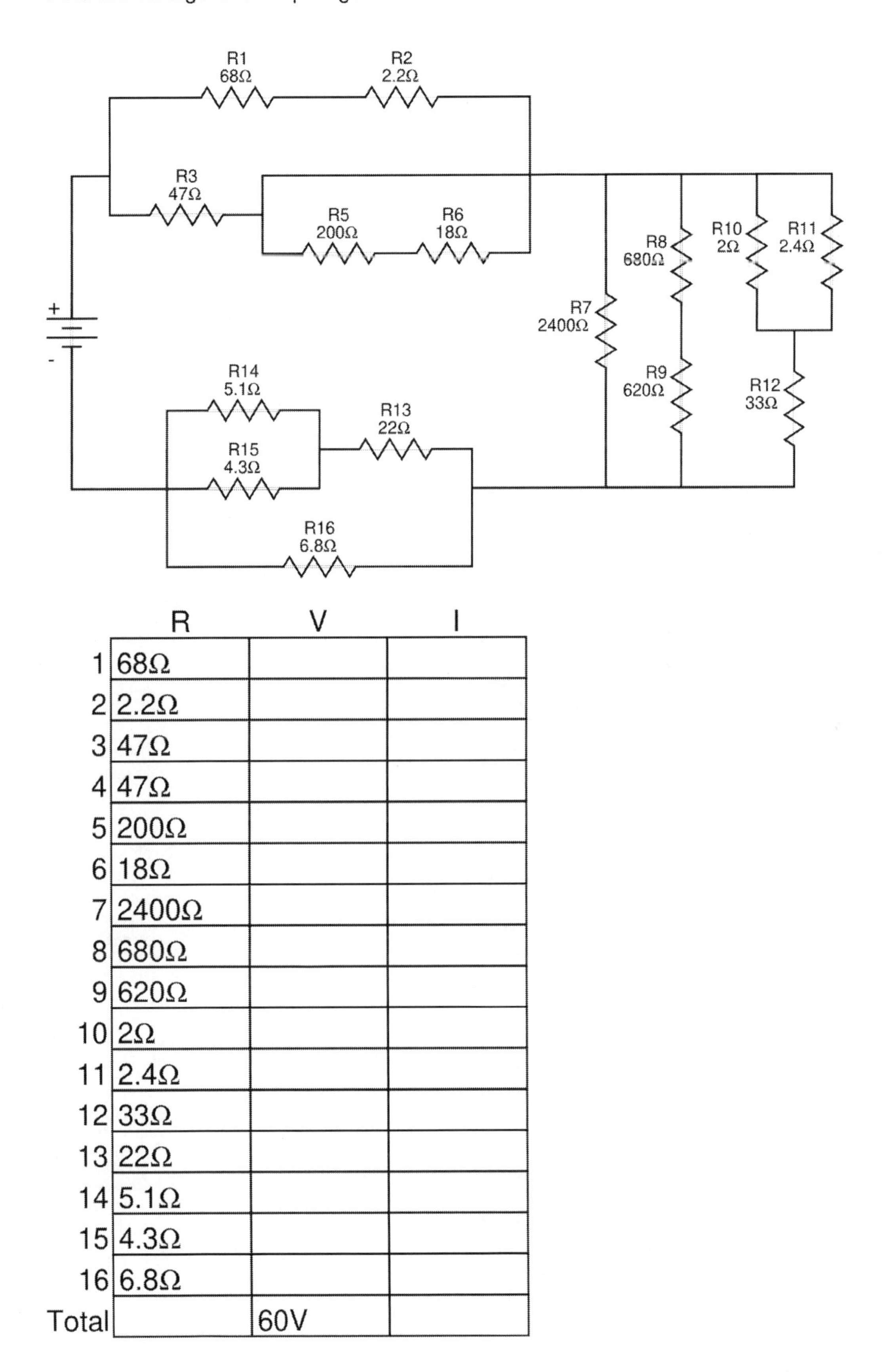

	R	V	I
1	68Ω		
2	2.2Ω		
3	47Ω		
4	47Ω		
5	200Ω		
6	18Ω		
7	2400Ω		
8	680Ω		
9	620Ω		
10	2Ω		
11	2.4Ω		
12	33Ω		
13	22Ω		
14	5.1Ω		
15	4.3Ω		
16	6.8Ω		
Total		60V	

Exercise #97

Find the voltage and amperage for each resistor.

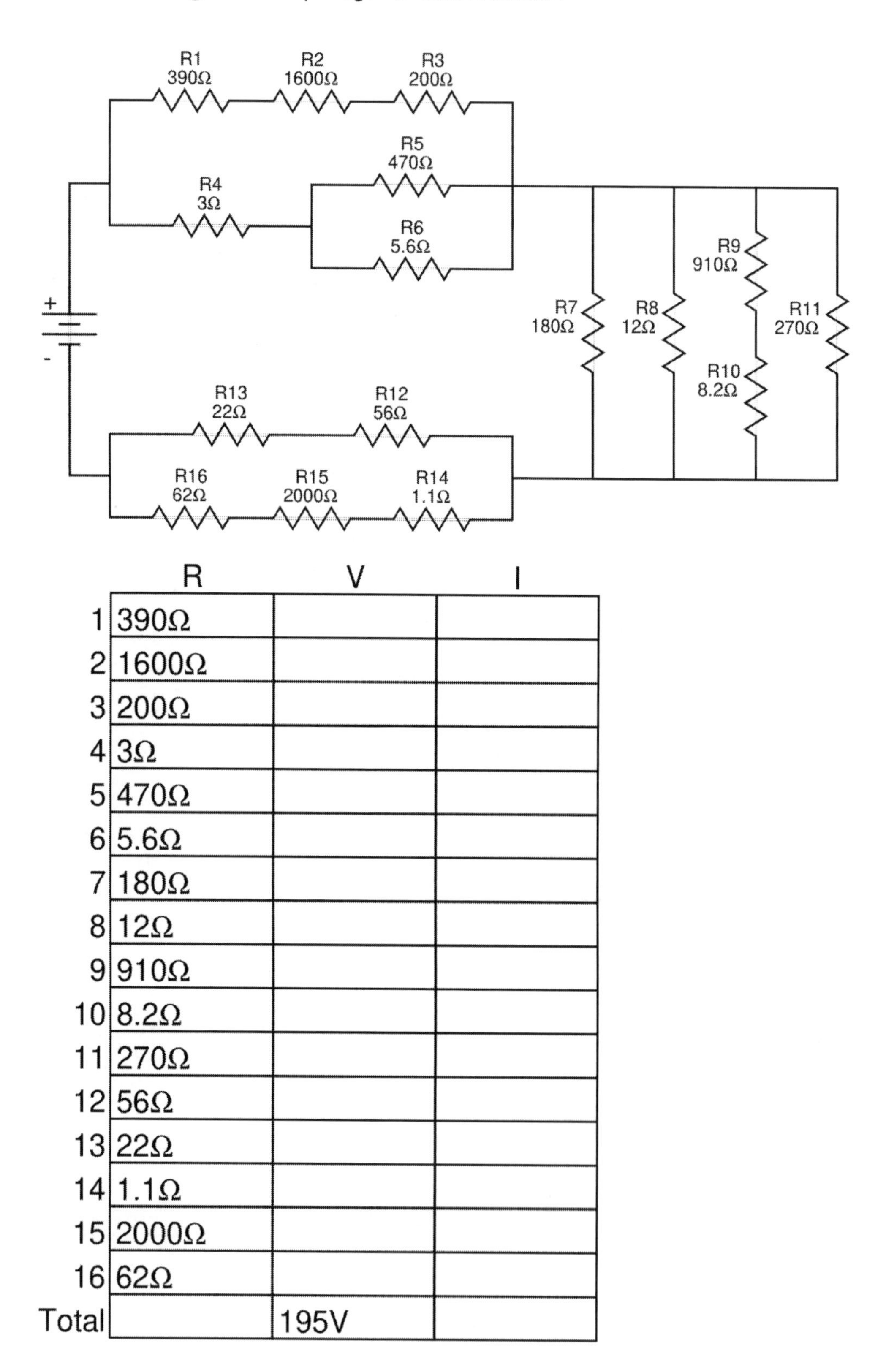

	R	V	I
1	390Ω		
2	1600Ω		
3	200Ω		
4	3Ω		
5	470Ω		
6	5.6Ω		
7	180Ω		
8	12Ω		
9	910Ω		
10	8.2Ω		
11	270Ω		
12	56Ω		
13	22Ω		
14	1.1Ω		
15	2000Ω		
16	62Ω		
Total		195V	

Exercise #98

Find the voltage and amperage for each resistor.

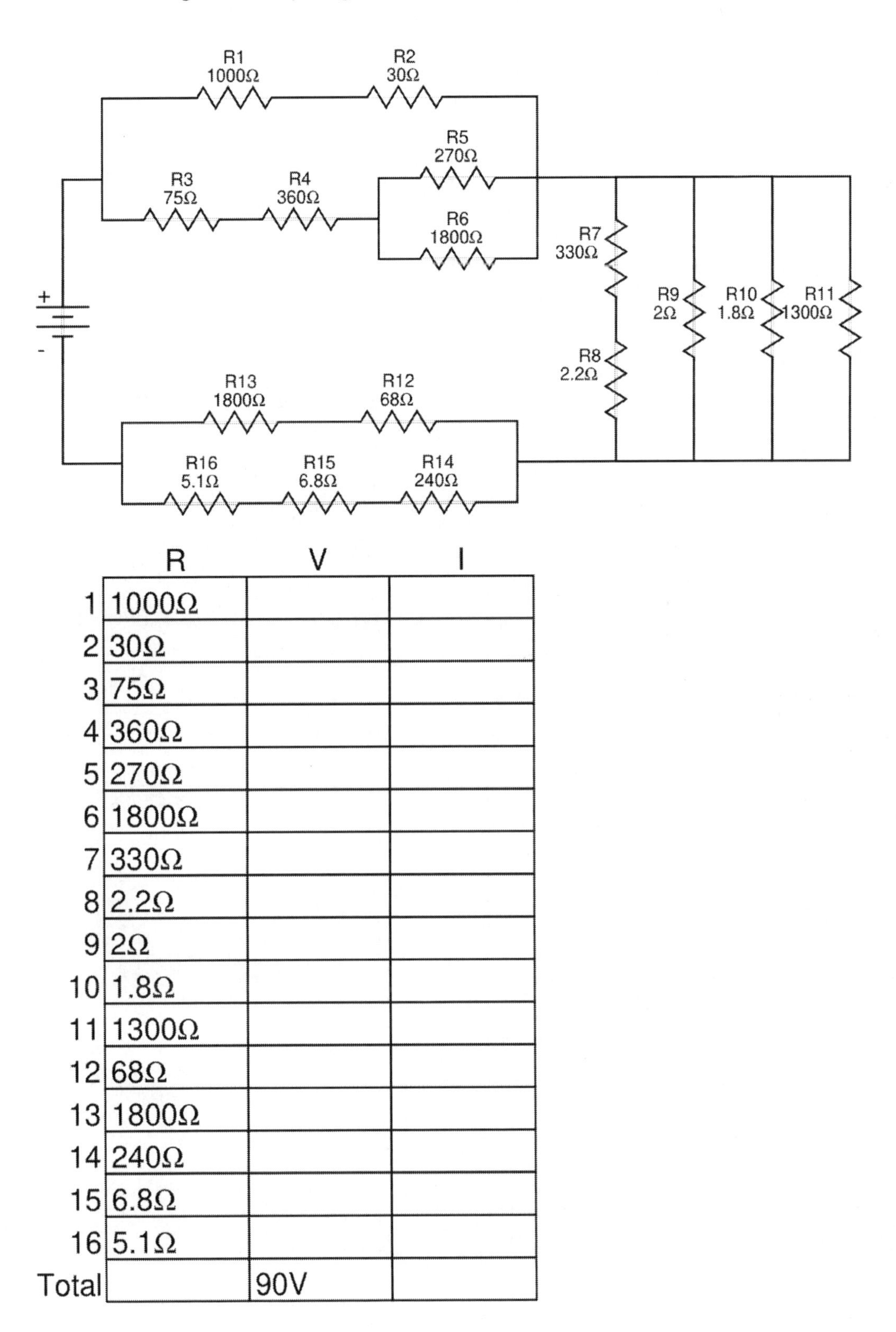

	R	V	I
1	1000Ω		
2	30Ω		
3	75Ω		
4	360Ω		
5	270Ω		
6	1800Ω		
7	330Ω		
8	2.2Ω		
9	2Ω		
10	1.8Ω		
11	1300Ω		
12	68Ω		
13	1800Ω		
14	240Ω		
15	6.8Ω		
16	5.1Ω		
Total		90V	

Exercise #99

Find the voltage and amperage for each resistor.

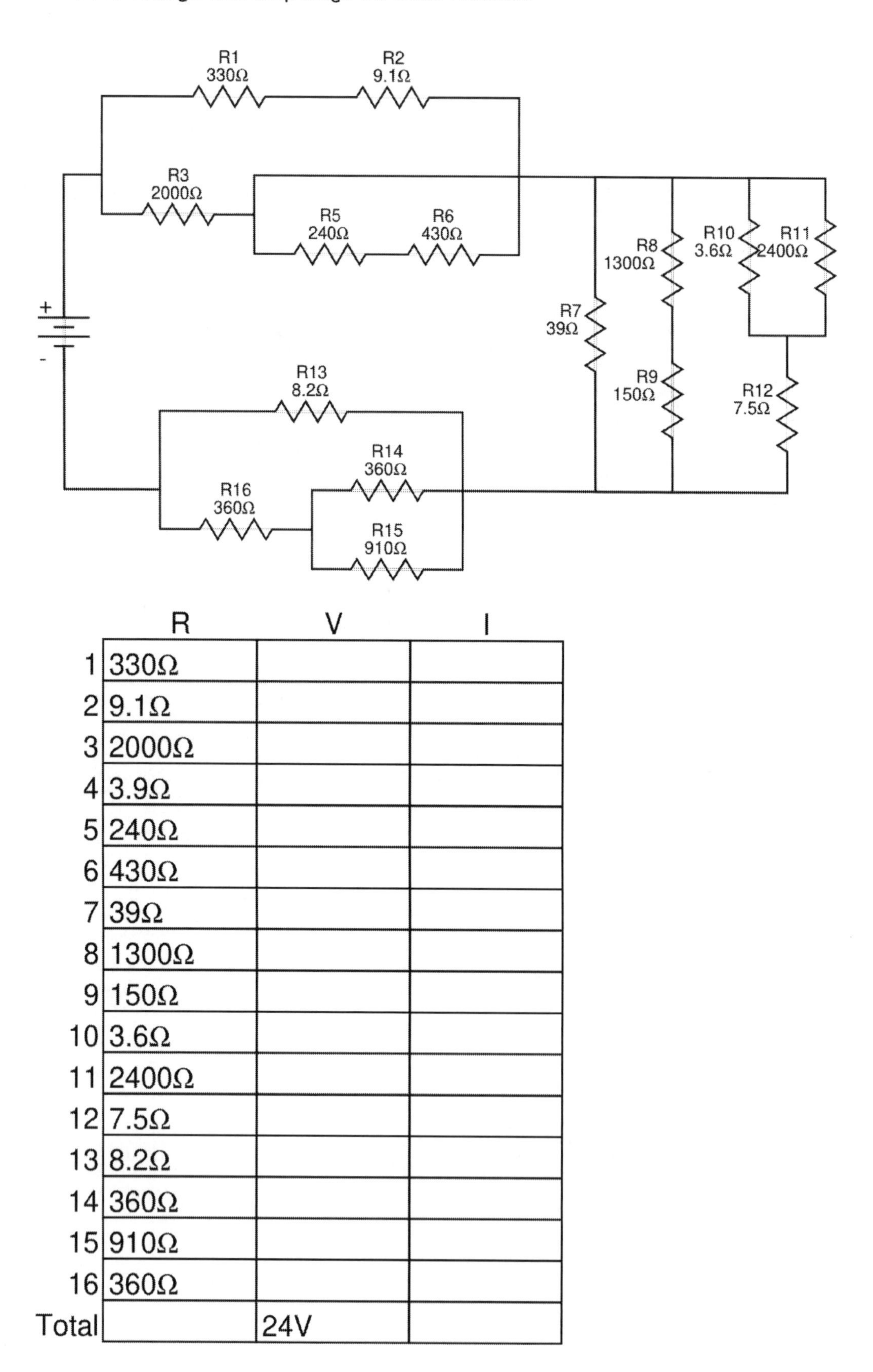

	R	V	I
1	330Ω		
2	9.1Ω		
3	2000Ω		
4	3.9Ω		
5	240Ω		
6	430Ω		
7	39Ω		
8	1300Ω		
9	150Ω		
10	3.6Ω		
11	2400Ω		
12	7.5Ω		
13	8.2Ω		
14	360Ω		
15	910Ω		
16	360Ω		
Total		24V	

Exercise #100

Find the voltage and amperage for each resistor.

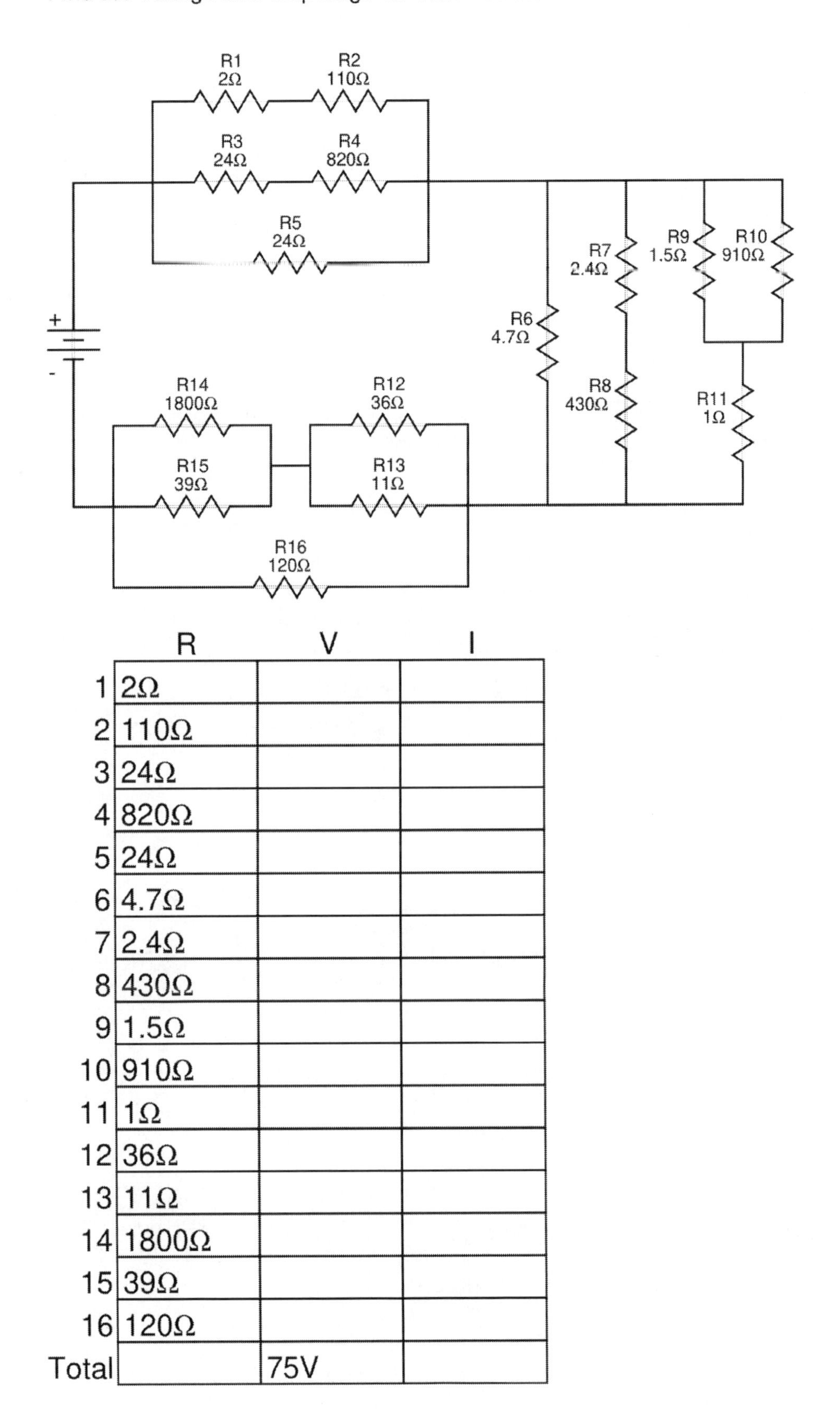

	R	V	I
1	2Ω		
2	110Ω		
3	24Ω		
4	820Ω		
5	24Ω		
6	4.7Ω		
7	2.4Ω		
8	430Ω		
9	1.5Ω		
10	910Ω		
11	1Ω		
12	36Ω		
13	11Ω		
14	1800Ω		
15	39Ω		
16	120Ω		
Total		75V	

Exercise #1

Find the voltage and amperage for each resistor.

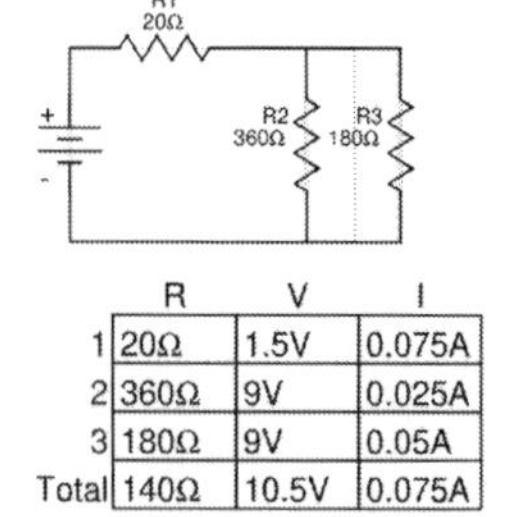

	R	V	I
1	20Ω	1.5V	0.075A
2	360Ω	9V	0.025A
3	180Ω	9V	0.05A
Total	140Ω	10.5V	0.075A

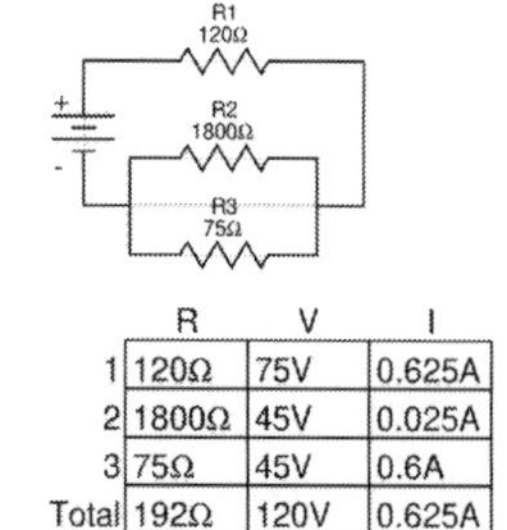

	R	V	I
1	120Ω	75V	0.625A
2	1800Ω	45V	0.025A
3	75Ω	45V	0.6A
Total	192Ω	120V	0.625A

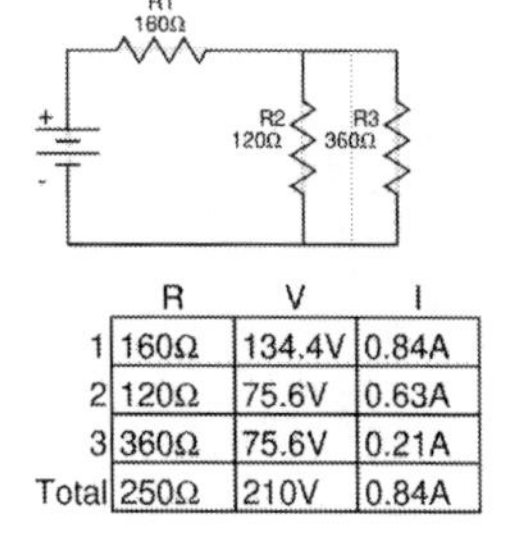

	R	V	I
1	160Ω	134.4V	0.84A
2	120Ω	75.6V	0.63A
3	360Ω	75.6V	0.21A
Total	250Ω	210V	0.84A

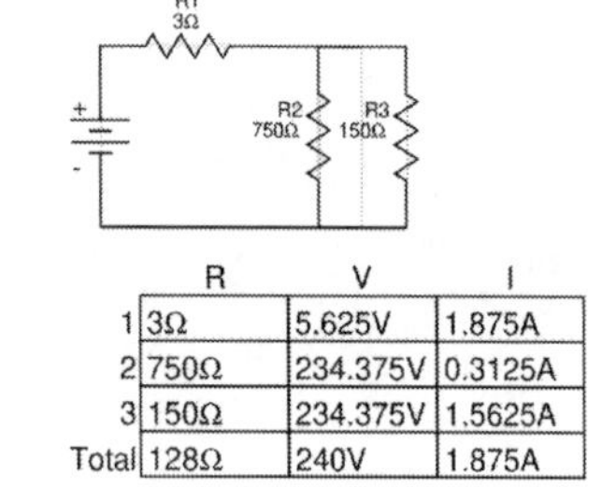

	R	V	I
1	3Ω	5.625V	1.875A
2	750Ω	234.375V	0.3125A
3	150Ω	234.375V	1.5625A
Total	128Ω	240V	1.875A

Exercise #2

Find the voltage and amperage for each resistor.

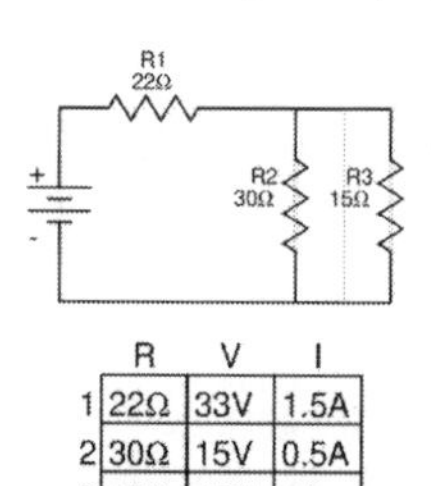

	R	V	I
1	22Ω	33V	1.5A
2	30Ω	15V	0.5A
3	15Ω	15V	1A
Total	32Ω	48V	1.5A

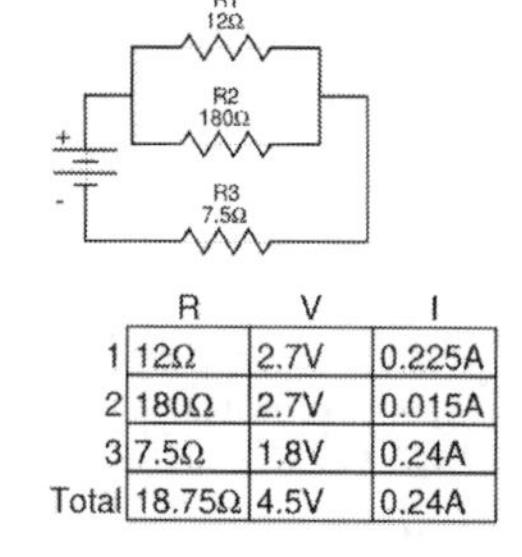

	R	V	I
1	12Ω	2.7V	0.225A
2	180Ω	2.7V	0.015A
3	7.5Ω	1.8V	0.24A
Total	18.75Ω	4.5V	0.24A

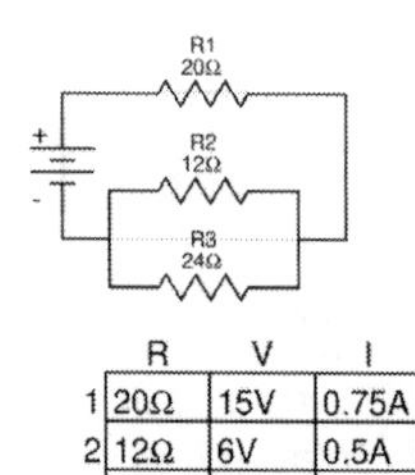

	R	V	I
1	20Ω	15V	0.75A
2	12Ω	6V	0.5A
3	24Ω	6V	0.25A
Total	28Ω	21V	0.75A

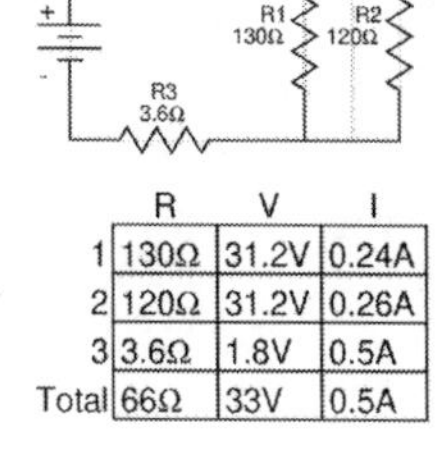

	R	V	I
1	130Ω	31.2V	0.24A
2	120Ω	31.2V	0.26A
3	3.6Ω	1.8V	0.5A
Total	66Ω	33V	0.5A

Exercise #3

Find the voltage and amperage for each resistor.

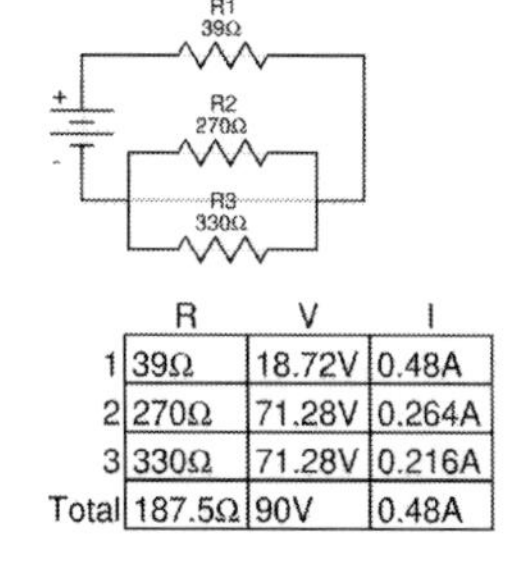

	R	V	I
1	39Ω	18.72V	0.48A
2	270Ω	71.28V	0.264A
3	330Ω	71.28V	0.216A
Total	187.5Ω	90V	0.48A

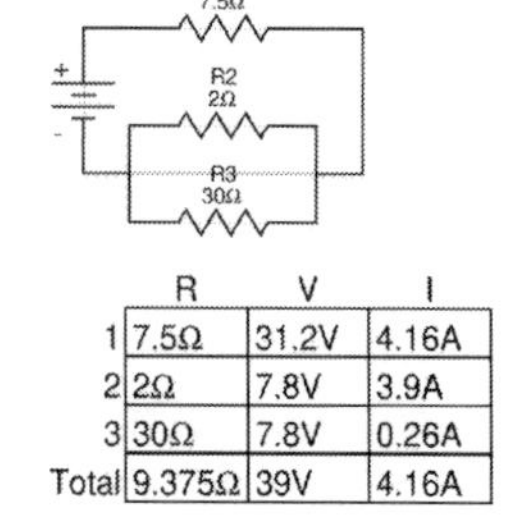

	R	V	I
1	7.5Ω	31.2V	4.16A
2	2Ω	7.8V	3.9A
3	30Ω	7.8V	0.26A
Total	9.375Ω	39V	4.16A

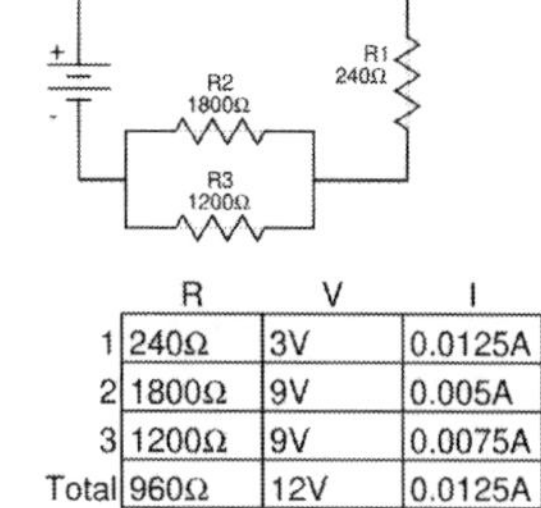

	R	V	I
1	240Ω	3V	0.0125A
2	1800Ω	9V	0.005A
3	1200Ω	9V	0.0075A
Total	960Ω	12V	0.0125A

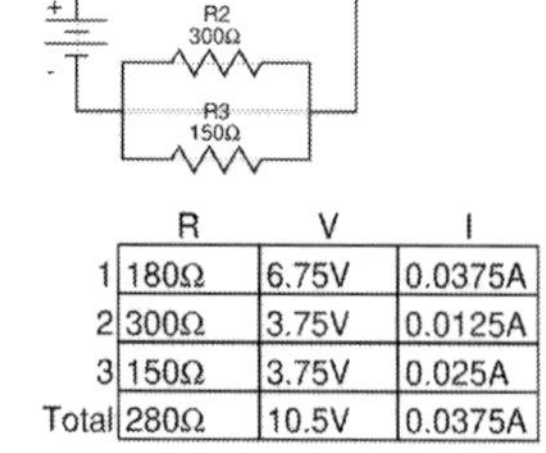

	R	V	I
1	180Ω	6.75V	0.0375A
2	300Ω	3.75V	0.0125A
3	150Ω	3.75V	0.025A
Total	280Ω	10.5V	0.0375A

Exercise #4

Find the voltage and amperage for each resistor.

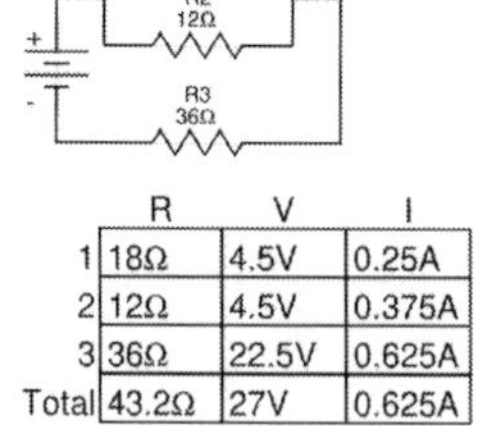

	R	V	I
1	18Ω	4.5V	0.25A
2	12Ω	4.5V	0.375A
3	36Ω	22.5V	0.625A
Total	43.2Ω	27V	0.625A

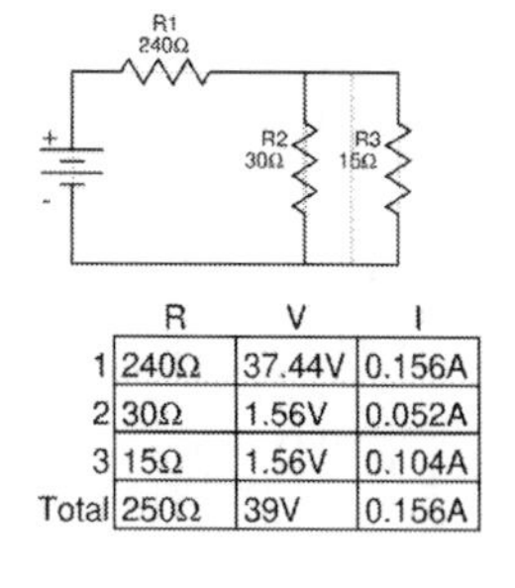

	R	V	I
1	240Ω	37.44V	0.156A
2	30Ω	1.56V	0.052A
3	15Ω	1.56V	0.104A
Total	250Ω	39V	0.156A

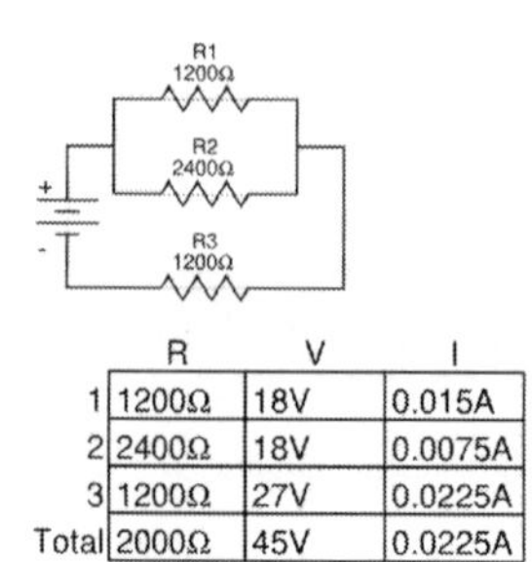

	R	V	I
1	1200Ω	18V	0.015A
2	2400Ω	18V	0.0075A
3	1200Ω	27V	0.0225A
Total	2000Ω	45V	0.0225A

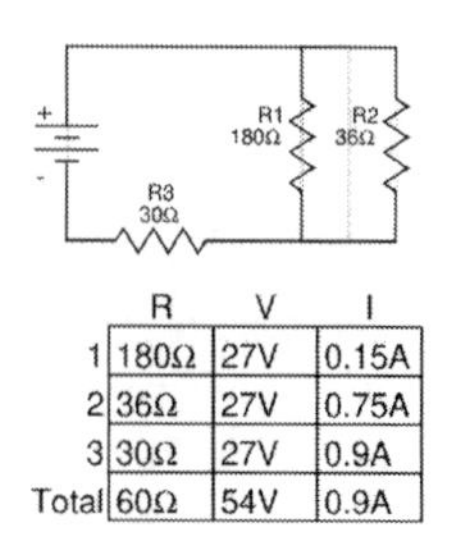

	R	V	I
1	180Ω	27V	0.15A
2	36Ω	27V	0.75A
3	30Ω	27V	0.9A
Total	60Ω	54V	0.9A

Exercise #5

Find the voltage and amperage for each resistor.

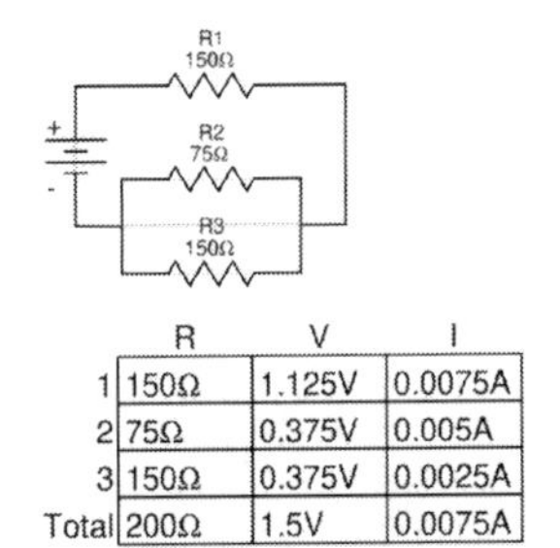

	R	V	I
1	150Ω	1.125V	0.0075A
2	75Ω	0.375V	0.005A
3	150Ω	0.375V	0.0025A
Total	200Ω	1.5V	0.0075A

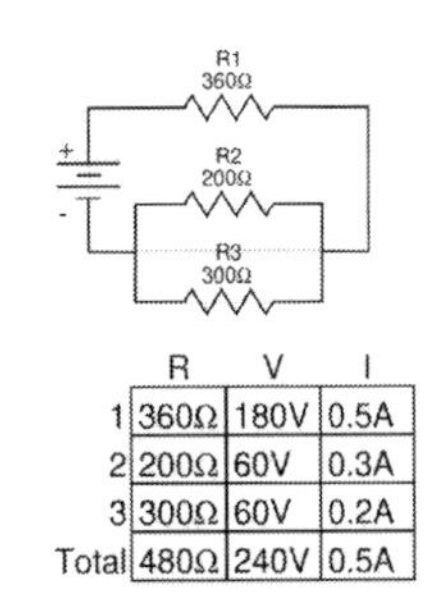

	R	V	I
1	360Ω	180V	0.5A
2	200Ω	60V	0.3A
3	300Ω	60V	0.2A
Total	480Ω	240V	0.5A

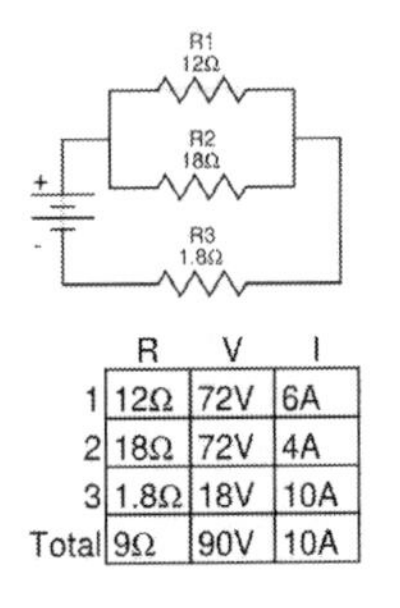

	R	V	I
1	12Ω	72V	6A
2	18Ω	72V	4A
3	1.8Ω	18V	10A
Total	9Ω	90V	10A

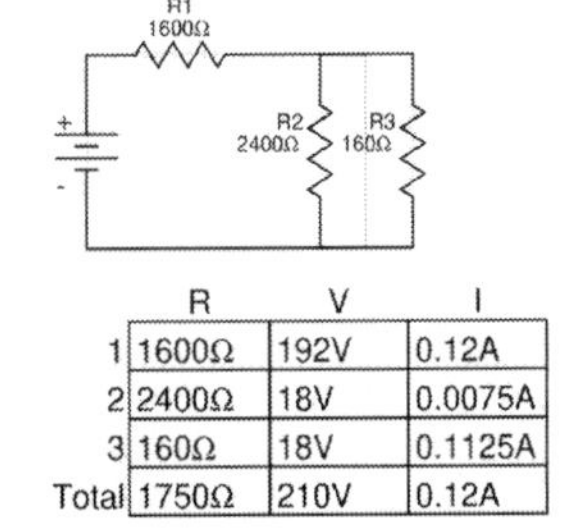

	R	V	I
1	1600Ω	192V	0.12A
2	2400Ω	18V	0.0075A
3	160Ω	18V	0.1125A
Total	1750Ω	210V	0.12A

Exercise #6

Find the voltage and amperage for each resistor.

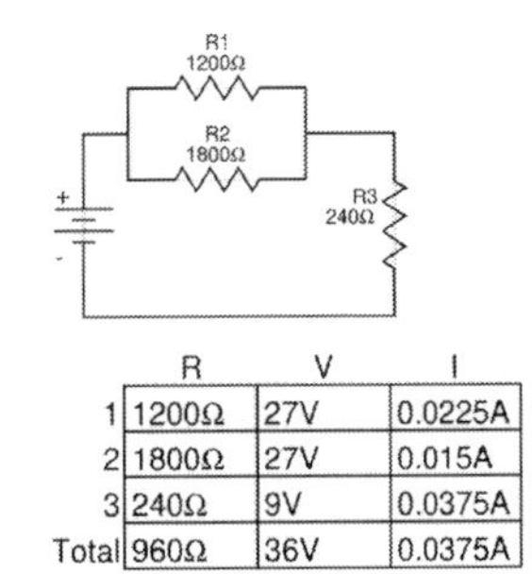

	R	V	I
1	1200Ω	27V	0.0225A
2	1800Ω	27V	0.015A
3	240Ω	9V	0.0375A
Total	960Ω	36V	0.0375A

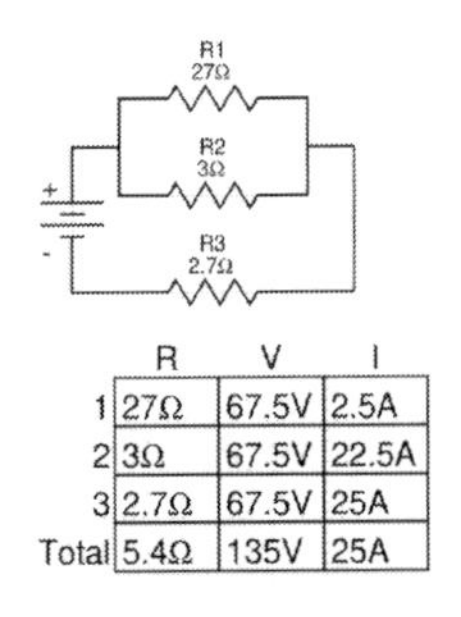

	R	V	I
1	27Ω	67.5V	2.5A
2	3Ω	67.5V	22.5A
3	2.7Ω	67.5V	25A
Total	5.4Ω	135V	25A

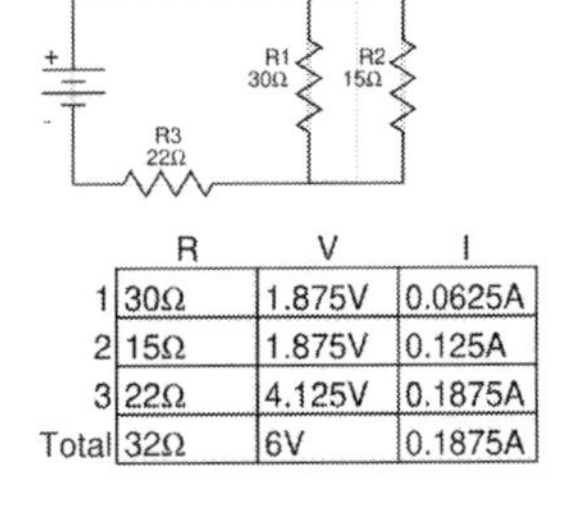

	R	V	I
1	30Ω	1.875V	0.0625A
2	15Ω	1.875V	0.125A
3	22Ω	4.125V	0.1875A
Total	32Ω	6V	0.1875A

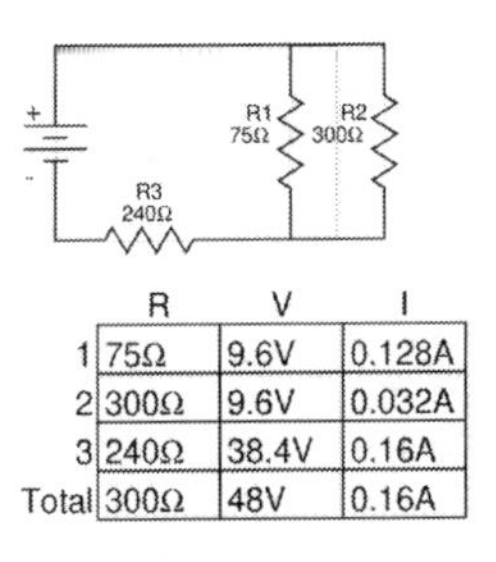

	R	V	I
1	75Ω	9.6V	0.128A
2	300Ω	9.6V	0.032A
3	240Ω	38.4V	0.16A
Total	300Ω	48V	0.16A

Exercise #7

Find the voltage and amperage for each resistor.

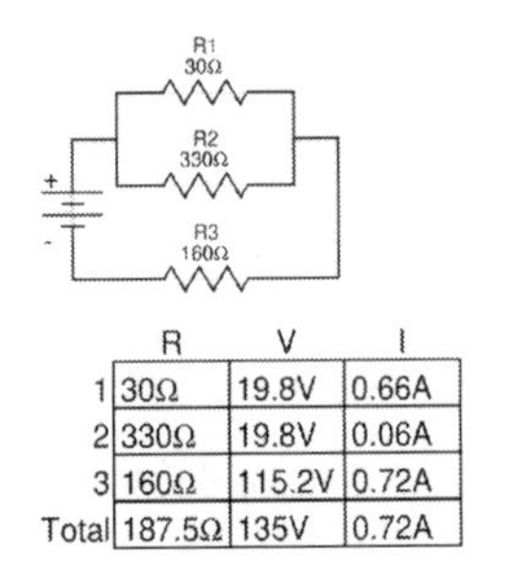

	R	V	I
1	30Ω	19.8V	0.66A
2	330Ω	19.8V	0.06A
3	160Ω	115.2V	0.72A
Total	187.5Ω	135V	0.72A

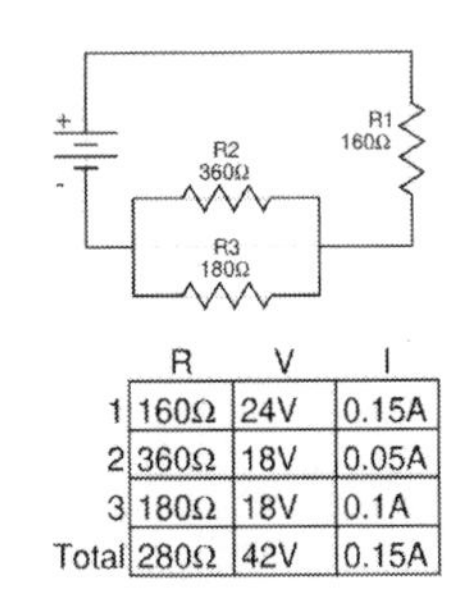

	R	V	I
1	160Ω	24V	0.15A
2	360Ω	18V	0.05A
3	180Ω	18V	0.1A
Total	280Ω	42V	0.15A

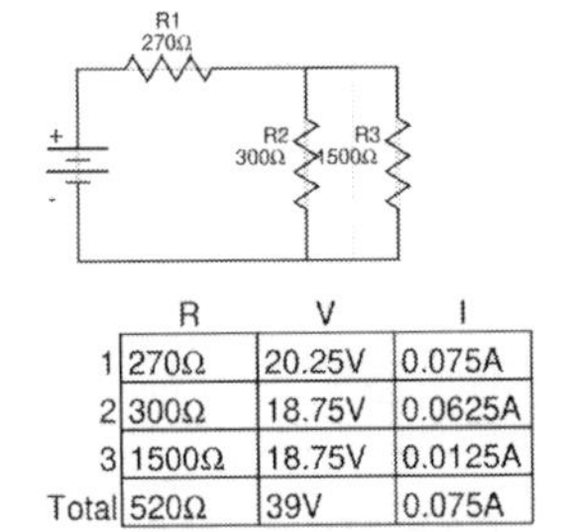

	R	V	I
1	270Ω	20.25V	0.075A
2	300Ω	18.75V	0.0625A
3	1500Ω	18.75V	0.0125A
Total	520Ω	39V	0.075A

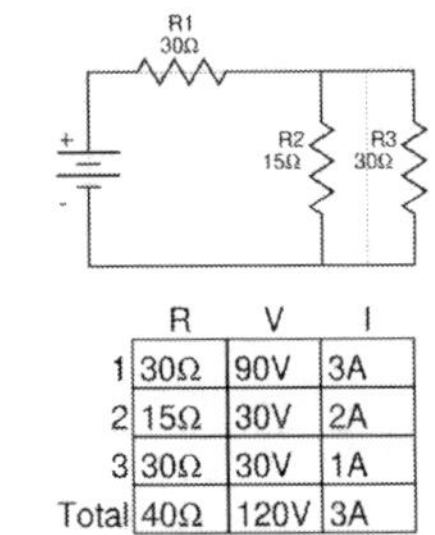

	R	V	I
1	30Ω	90V	3A
2	15Ω	30V	2A
3	30Ω	30V	1A
Total	40Ω	120V	3A

Exercise #8

Find the voltage and amperage for each resistor.

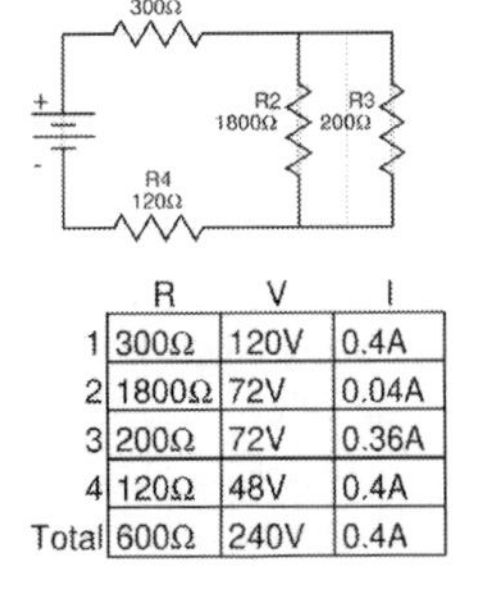

	R	V	I
1	300Ω	120V	0.4A
2	1800Ω	72V	0.04A
3	200Ω	72V	0.36A
4	120Ω	48V	0.4A
Total	600Ω	240V	0.4A

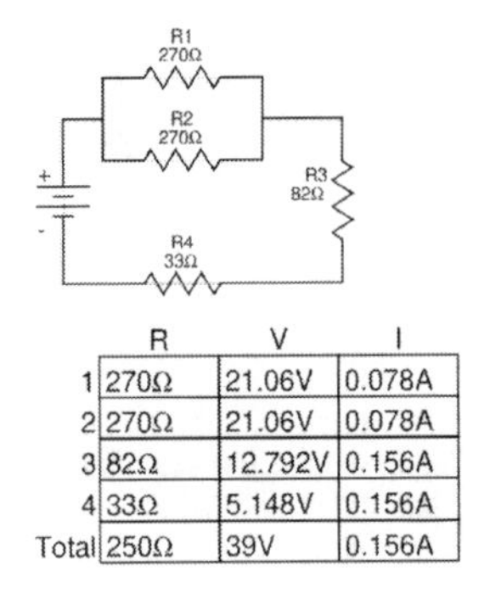

	R	V	I
1	270Ω	21.06V	0.078A
2	270Ω	21.06V	0.078A
3	82Ω	12.792V	0.156A
4	33Ω	5.148V	0.156A
Total	250Ω	39V	0.156A

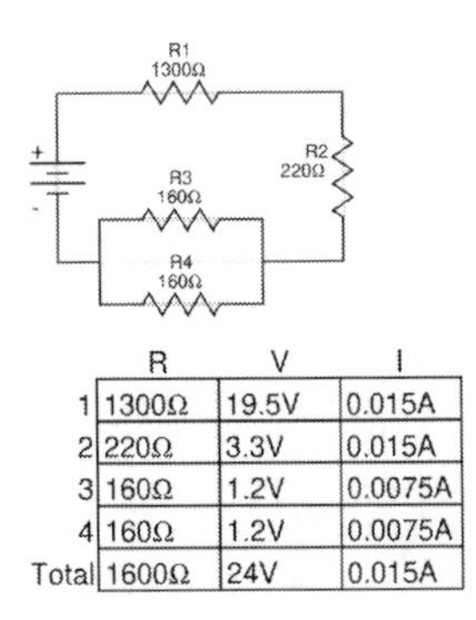

	R	V	I
1	1300Ω	19.5V	0.015A
2	220Ω	3.3V	0.015A
3	160Ω	1.2V	0.0075A
4	160Ω	1.2V	0.0075A
Total	1600Ω	24V	0.015A

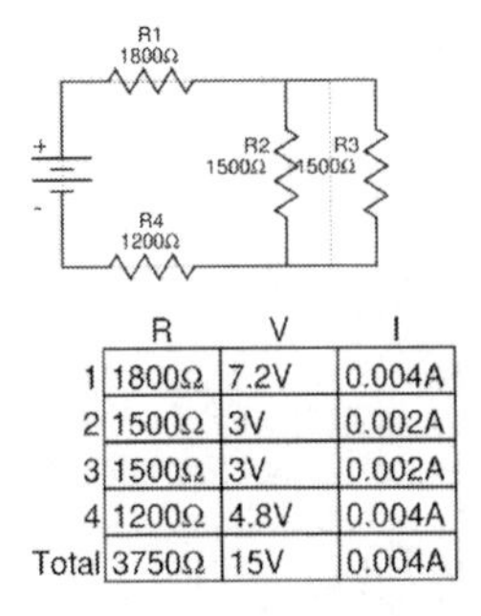

	R	V	I
1	1800Ω	7.2V	0.004A
2	1500Ω	3V	0.002A
3	1500Ω	3V	0.002A
4	1200Ω	4.8V	0.004A
Total	3750Ω	15V	0.004A

Exercise #9

Find the voltage and amperage for each resistor.

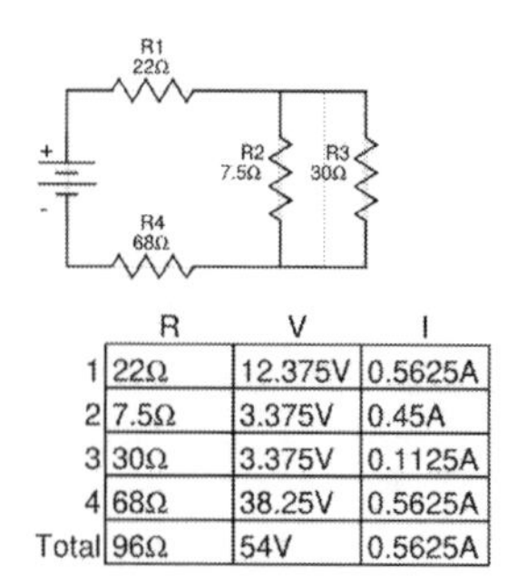

	R	V	I
1	22Ω	12.375V	0.5625A
2	7.5Ω	3.375V	0.45A
3	30Ω	3.375V	0.1125A
4	68Ω	38.25V	0.5625A
Total	96Ω	54V	0.5625A

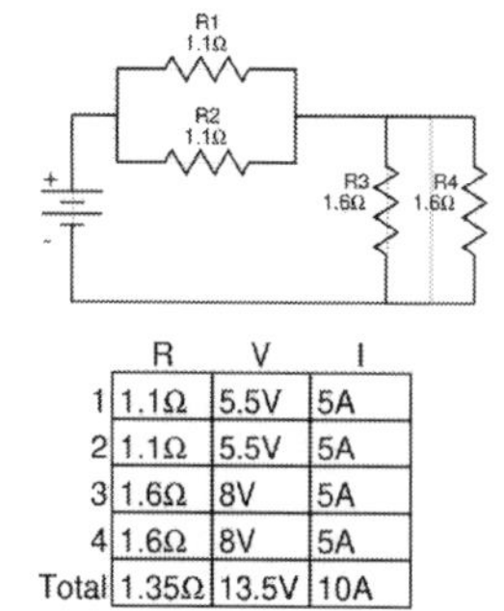

	R	V	I
1	1.1Ω	5.5V	5A
2	1.1Ω	5.5V	5A
3	1.6Ω	8V	5A
4	1.6Ω	8V	5A
Total	1.35Ω	13.5V	10A

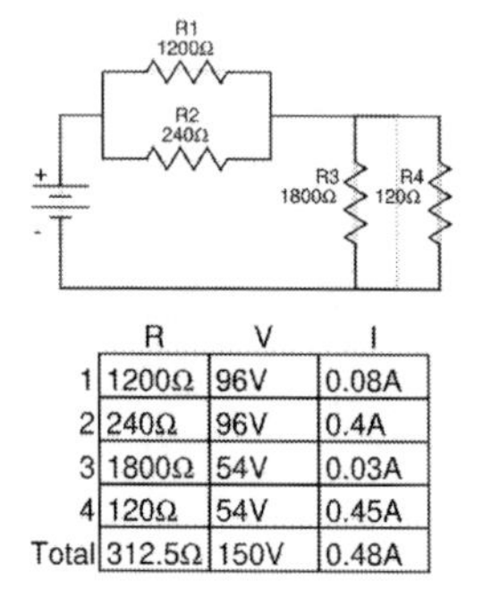

	R	V	I
1	1200Ω	96V	0.08A
2	240Ω	96V	0.4A
3	1800Ω	54V	0.03A
4	120Ω	54V	0.45A
Total	312.5Ω	150V	0.48A

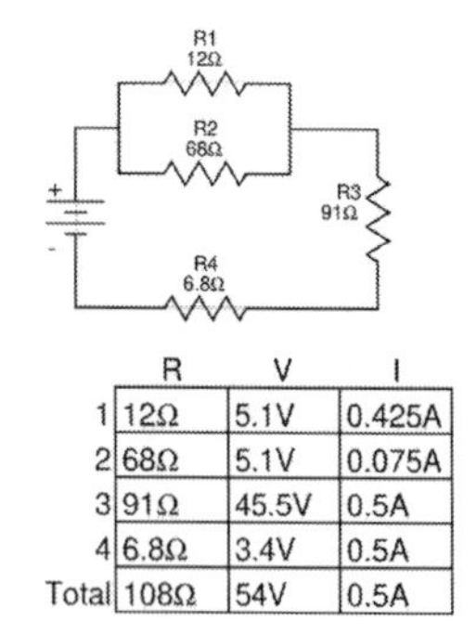

	R	V	I
1	12Ω	5.1V	0.425A
2	68Ω	5.1V	0.075A
3	91Ω	45.5V	0.5A
4	6.8Ω	3.4V	0.5A
Total	108Ω	54V	0.5A

Exercise #10

Find the voltage and amperage for each resistor.

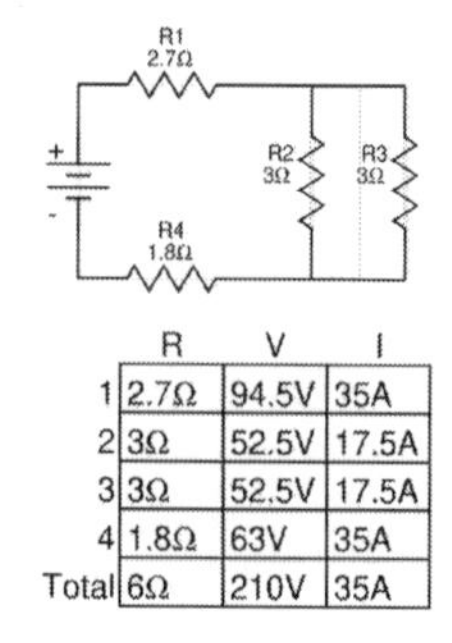

	R	V	I
1	2.7Ω	94.5V	35A
2	3Ω	52.5V	17.5A
3	3Ω	52.5V	17.5A
4	1.8Ω	63V	35A
Total	6Ω	210V	35A

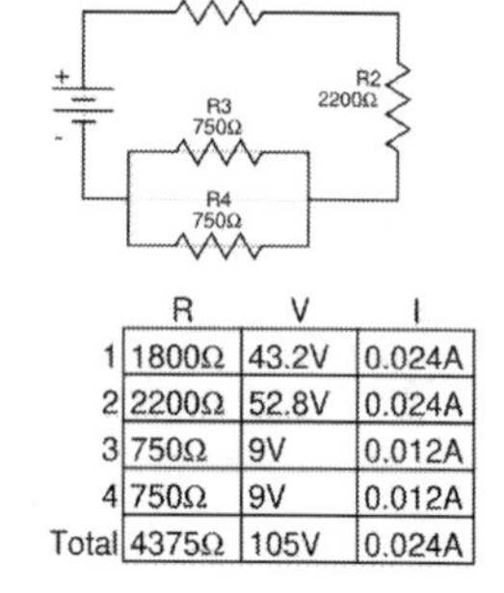

	R	V	I
1	1800Ω	43.2V	0.024A
2	2200Ω	52.8V	0.024A
3	750Ω	9V	0.012A
4	750Ω	9V	0.012A
Total	4375Ω	105V	0.024A

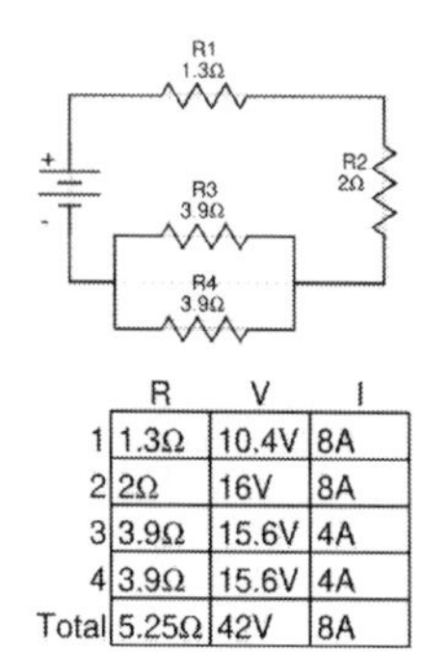

	R	V	I
1	1.3Ω	10.4V	8A
2	2Ω	16V	8A
3	3.9Ω	15.6V	4A
4	3.9Ω	15.6V	4A
Total	5.25Ω	42V	8A

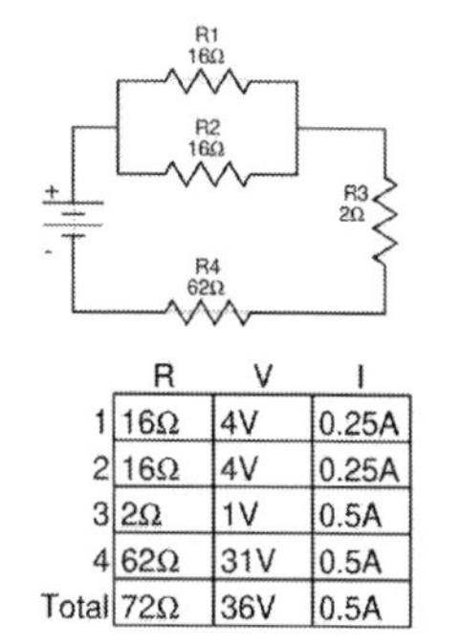

	R	V	I
1	16Ω	4V	0.25A
2	16Ω	4V	0.25A
3	2Ω	1V	0.5A
4	62Ω	31V	0.5A
Total	72Ω	36V	0.5A

Exercise #11

Find the voltage and amperage for each resistor.

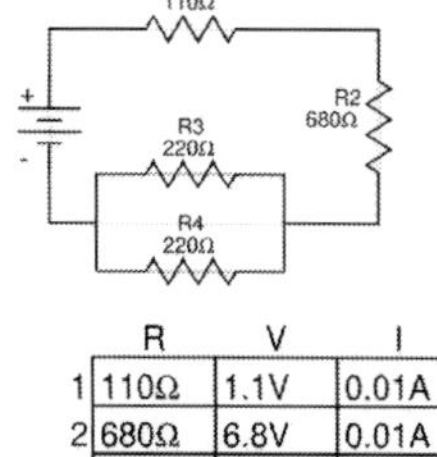

	R	V	I
1	110Ω	1.1V	0.01A
2	680Ω	6.8V	0.01A
3	220Ω	1.1V	0.005A
4	220Ω	1.1V	0.005A
Total	900Ω	9V	0.01A

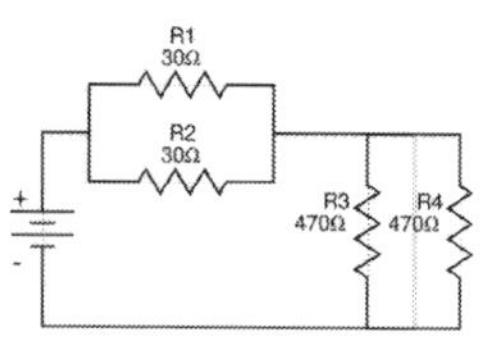

	R	V	I
1	30Ω	4.5V	0.15A
2	30Ω	4.5V	0.15A
3	470Ω	70.5V	0.15A
4	470Ω	70.5V	0.15A
Total	250Ω	75V	0.3A

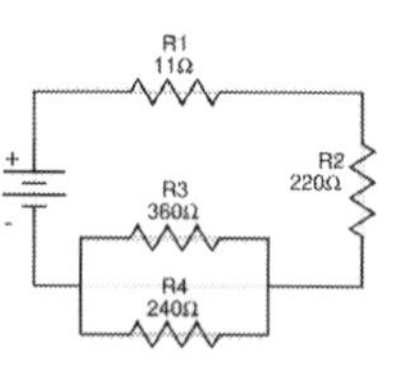

	R	V	I
1	11Ω	1.76V	0.16A
2	220Ω	35.2V	0.16A
3	360Ω	23.04V	0.064A
4	240Ω	23.04V	0.096A
Total	375Ω	60V	0.16A

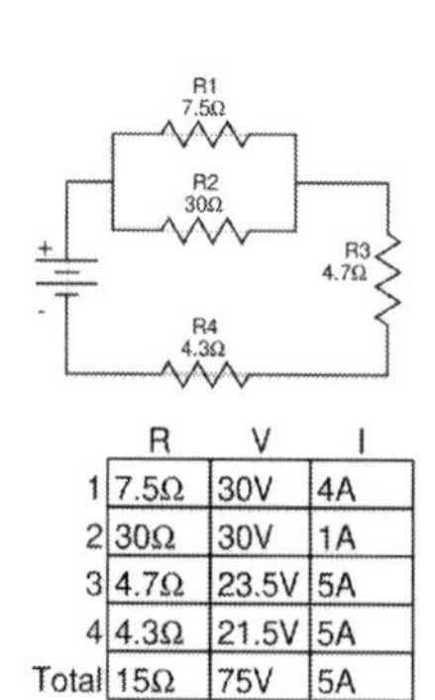

	R	V	I
1	7.5Ω	30V	4A
2	30Ω	30V	1A
3	4.7Ω	23.5V	5A
4	4.3Ω	21.5V	5A
Total	15Ω	75V	5A

Exercise #12

Find the voltage and amperage for each resistor.

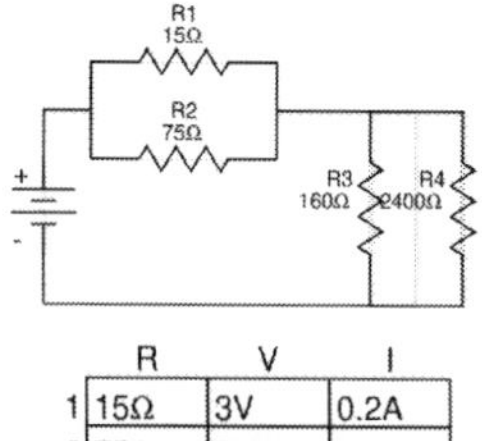

	R	V	I
1	15Ω	3V	0.2A
2	75Ω	3V	0.04A
3	160Ω	36V	0.225A
4	2400Ω	36V	0.015A
Total	162.5Ω	39V	0.24A

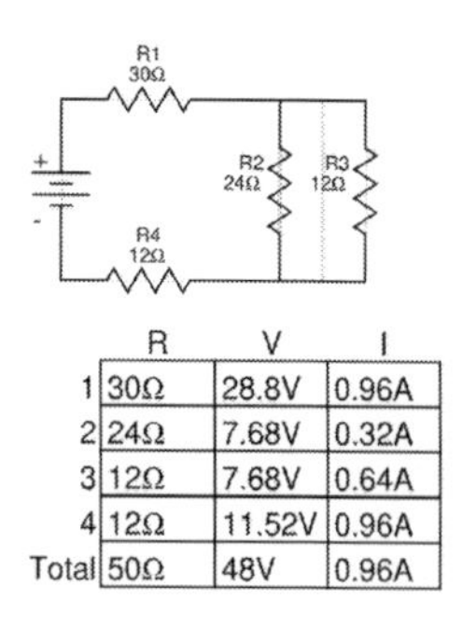

	R	V	I
1	30Ω	28.8V	0.96A
2	24Ω	7.68V	0.32A
3	12Ω	7.68V	0.64A
4	12Ω	11.52V	0.96A
Total	50Ω	48V	0.96A

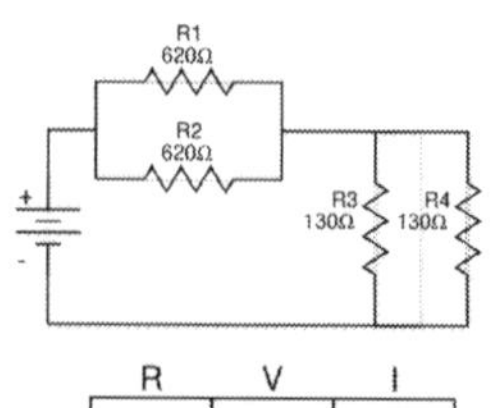

	R	V	I
1	620Ω	39.68V	0.064A
2	620Ω	39.68V	0.064A
3	130Ω	8.32V	0.064A
4	130Ω	8.32V	0.064A
Total	375Ω	48V	0.128A

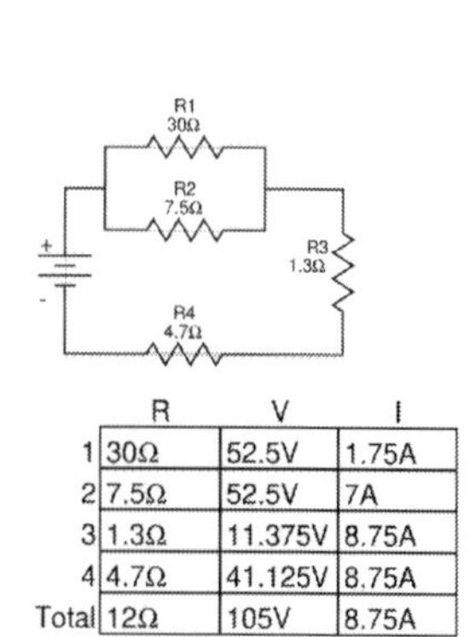

	R	V	I
1	30Ω	52.5V	1.75A
2	7.5Ω	52.5V	7A
3	1.3Ω	11.375V	8.75A
4	4.7Ω	41.125V	8.75A
Total	12Ω	105V	8.75A

Exercise #13

Find the voltage and amperage for each resistor.

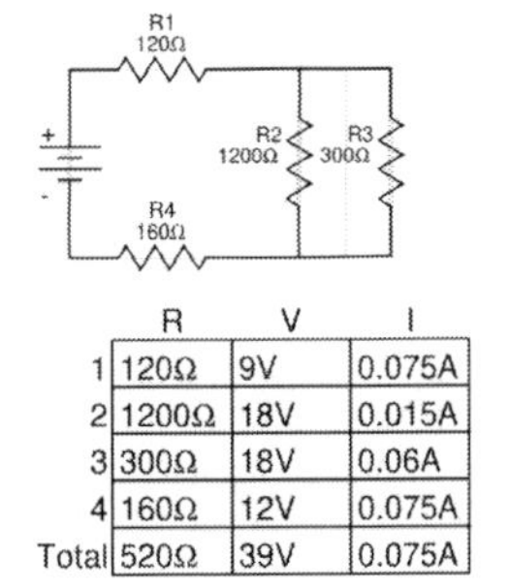

	R	V	I
1	120Ω	9V	0.075A
2	1200Ω	18V	0.015A
3	300Ω	18V	0.06A
4	160Ω	12V	0.075A
Total	520Ω	39V	0.075A

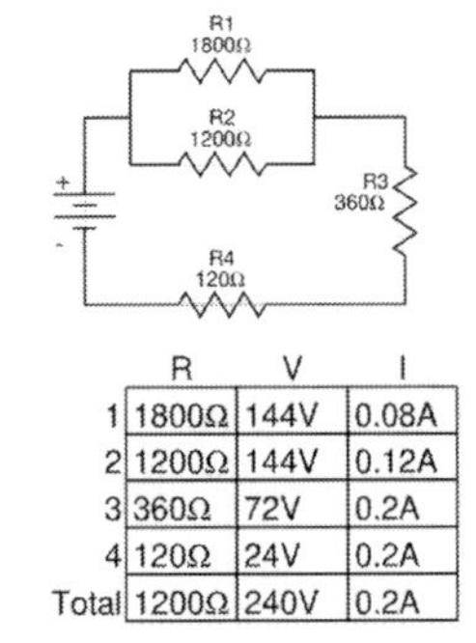

	R	V	I
1	1800Ω	144V	0.08A
2	1200Ω	144V	0.12A
3	360Ω	72V	0.2A
4	120Ω	24V	0.2A
Total	1200Ω	240V	0.2A

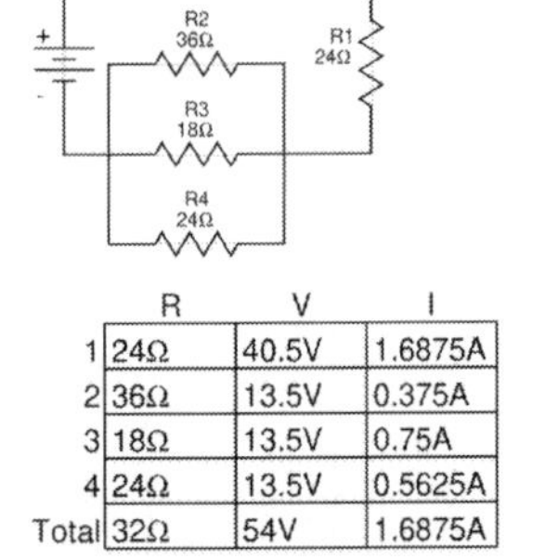

	R	V	I
1	24Ω	40.5V	1.6875A
2	36Ω	13.5V	0.375A
3	18Ω	13.5V	0.75A
4	24Ω	13.5V	0.5625A
Total	32Ω	54V	1.6875A

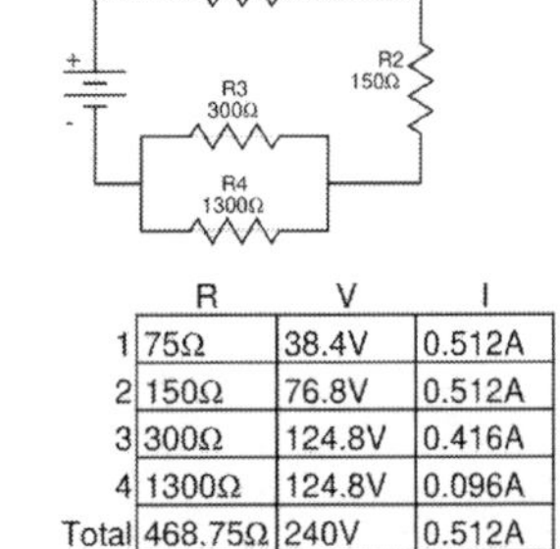

	R	V	I
1	75Ω	38.4V	0.512A
2	150Ω	76.8V	0.512A
3	300Ω	124.8V	0.416A
4	1300Ω	124.8V	0.096A
Total	468.75Ω	240V	0.512A

Exercise #14

Find the voltage and amperage for each resistor.

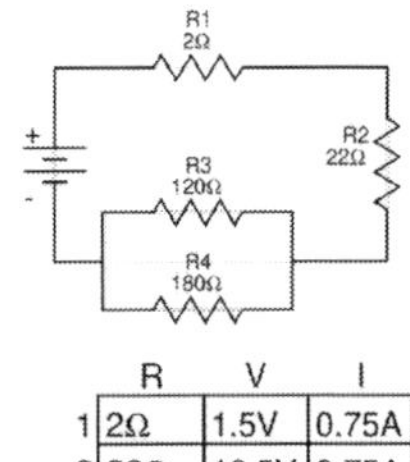

	R	V	I
1	2Ω	1.5V	0.75A
2	22Ω	16.5V	0.75A
3	120Ω	54V	0.45A
4	180Ω	54V	0.3A
Total	96Ω	72V	0.75A

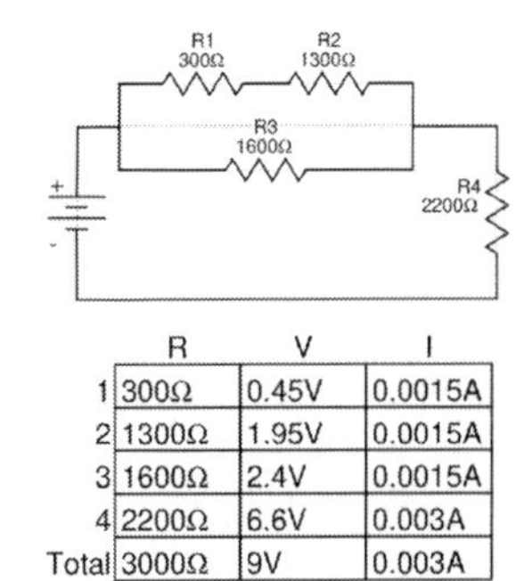

	R	V	I
1	300Ω	0.45V	0.0015A
2	1300Ω	1.95V	0.0015A
3	1600Ω	2.4V	0.0015A
4	2200Ω	6.6V	0.003A
Total	3000Ω	9V	0.003A

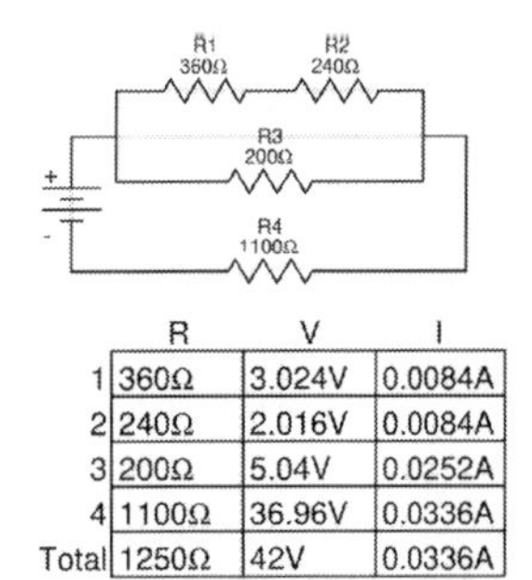

	R	V	I
1	360Ω	3.024V	0.0084A
2	240Ω	2.016V	0.0084A
3	200Ω	5.04V	0.0252A
4	1100Ω	36.96V	0.0336A
Total	1250Ω	42V	0.0336A

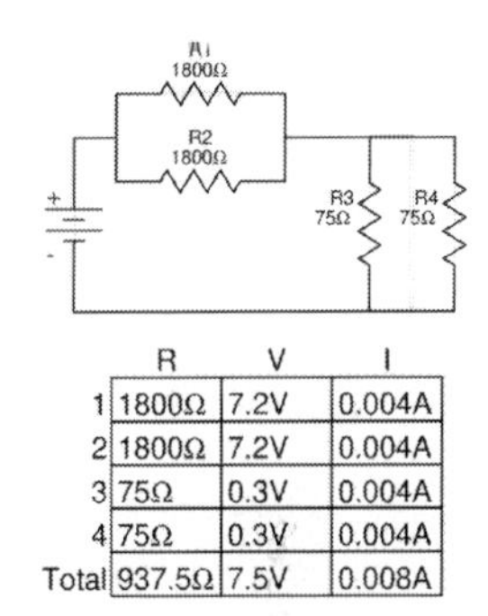

	R	V	I
1	1800Ω	7.2V	0.004A
2	1800Ω	7.2V	0.004A
3	75Ω	0.3V	0.004A
4	75Ω	0.3V	0.004A
Total	937.5Ω	7.5V	0.008A

Exercise #15

Find the voltage and amperage for each resistor.

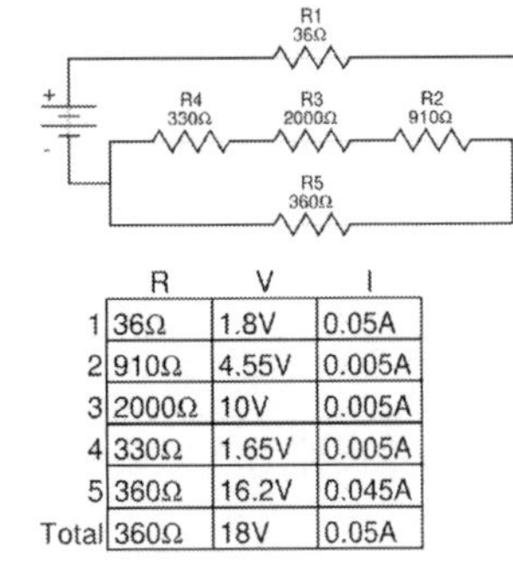

	R	V	I
1	36Ω	1.8V	0.05A
2	910Ω	4.55V	0.005A
3	2000Ω	10V	0.005A
4	330Ω	1.65V	0.005A
5	360Ω	16.2V	0.045A
Total	360Ω	18V	0.05A

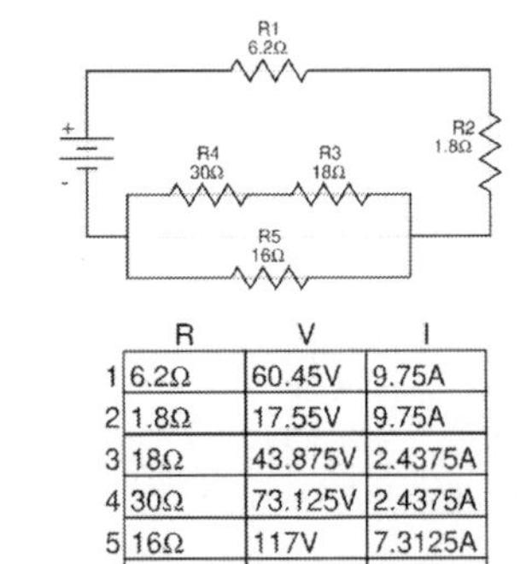

	R	V	I
1	6.2Ω	60.45V	9.75A
2	1.8Ω	17.55V	9.75A
3	18Ω	43.875V	2.4375A
4	30Ω	73.125V	2.4375A
5	16Ω	117V	7.3125A
Total	20Ω	195V	9.75A

Exercise #16

Find the voltage and amperage for each resistor.

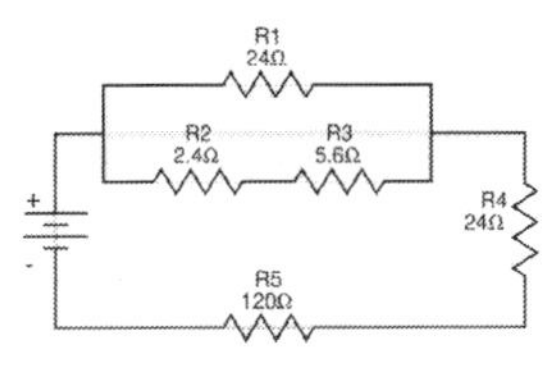

	R	V	I
1	24Ω	3.6V	0.15A
2	2.4Ω	1.08V	0.45A
3	5.6Ω	2.52V	0.45A
4	24Ω	14.4V	0.6A
5	120Ω	72V	0.6A
Total	150Ω	90V	0.6A

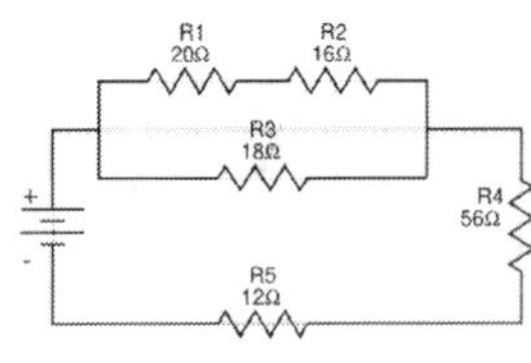

	R	V	I
1	20Ω	8.75V	0.4375A
2	16Ω	7V	0.4375A
3	18Ω	15.75V	0.875A
4	56Ω	73.5V	1.3125A
5	12Ω	15.75V	1.3125A
Total	80Ω	105V	1.3125A

Exercise #17

Find the voltage and amperage for each resistor.

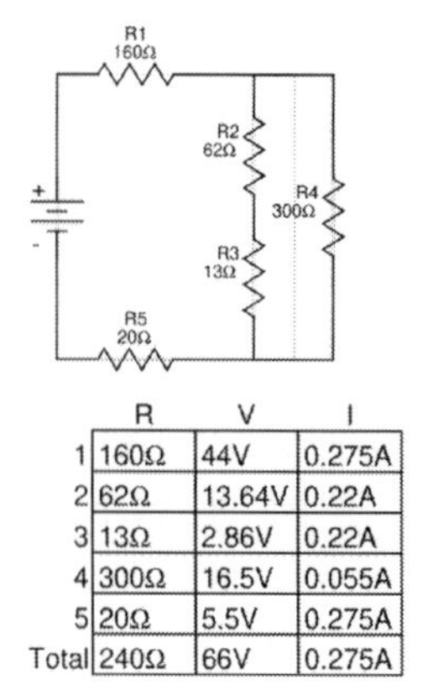

	R	V	I
1	160Ω	44V	0.275A
2	62Ω	13.64V	0.22A
3	13Ω	2.86V	0.22A
4	300Ω	16.5V	0.055A
5	20Ω	5.5V	0.275A
Total	240Ω	66V	0.275A

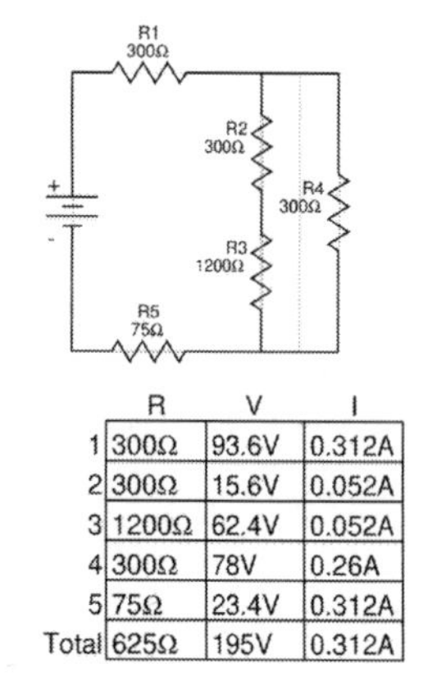

	R	V	I
1	300Ω	93.6V	0.312A
2	300Ω	15.6V	0.052A
3	1200Ω	62.4V	0.052A
4	300Ω	78V	0.26A
5	75Ω	23.4V	0.312A
Total	625Ω	195V	0.312A

Exercise #18

Find the voltage and amperage for each resistor.

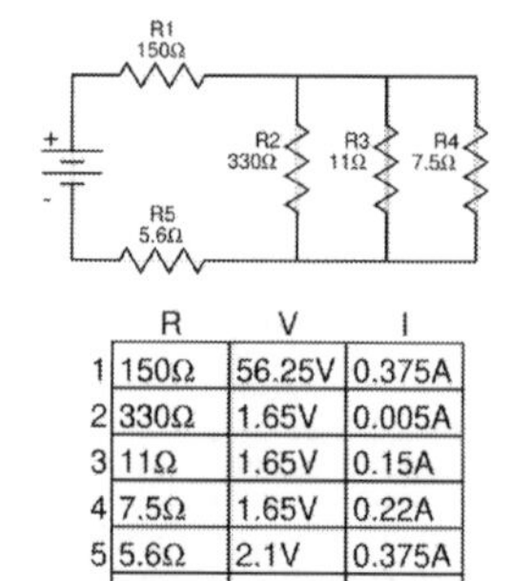

	R	V	I
1	150Ω	56.25V	0.375A
2	330Ω	1.65V	0.005A
3	11Ω	1.65V	0.15A
4	7.5Ω	1.65V	0.22A
5	5.6Ω	2.1V	0.375A
Total	160Ω	60V	0.375A

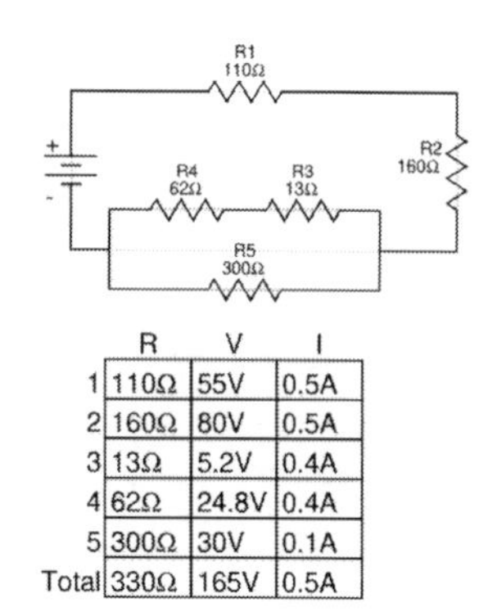

	R	V	I
1	110Ω	55V	0.5A
2	160Ω	80V	0.5A
3	13Ω	5.2V	0.4A
4	62Ω	24.8V	0.4A
5	300Ω	30V	0.1A
Total	330Ω	165V	0.5A

Exercise #19

Find the voltage and amperage for each resistor.

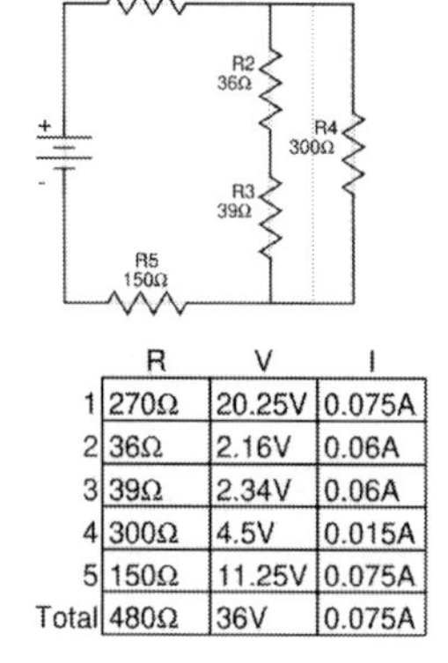

	R	V	I
1	270Ω	20.25V	0.075A
2	36Ω	2.16V	0.06A
3	39Ω	2.34V	0.06A
4	300Ω	4.5V	0.015A
5	150Ω	11.25V	0.075A
Total	480Ω	36V	0.075A

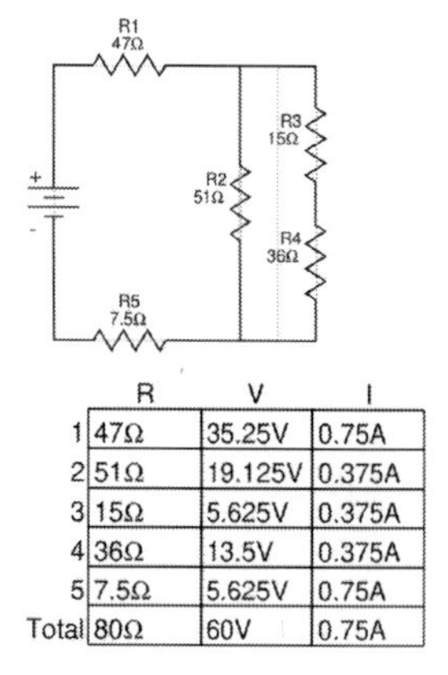

	R	V	I
1	47Ω	35.25V	0.75A
2	51Ω	19.125V	0.375A
3	15Ω	5.625V	0.375A
4	36Ω	13.5V	0.375A
5	7.5Ω	5.625V	0.75A
Total	80Ω	60V	0.75A

Exercise #20

Find the voltage and amperage for each resistor.

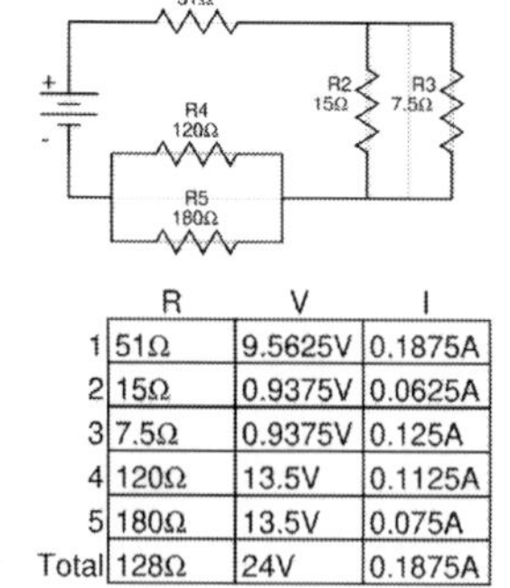

	R	V	I
1	51Ω	9.5625V	0.1875A
2	15Ω	0.9375V	0.0625A
3	7.5Ω	0.9375V	0.125A
4	120Ω	13.5V	0.1125A
5	180Ω	13.5V	0.075A
Total	128Ω	24V	0.1875A

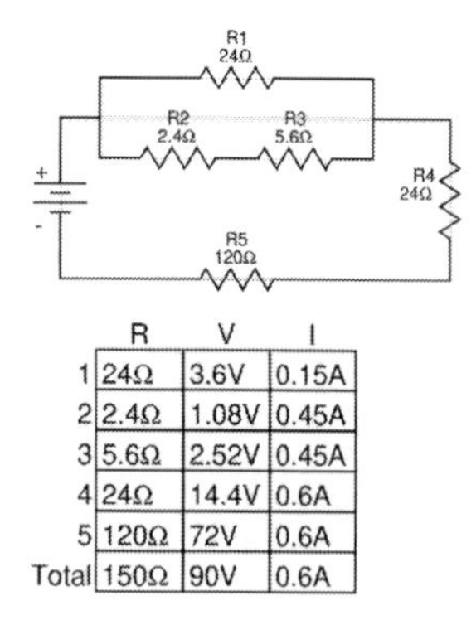

	R	V	I
1	24Ω	3.6V	0.15A
2	2.4Ω	1.08V	0.45A
3	5.6Ω	2.52V	0.45A
4	24Ω	14.4V	0.6A
5	120Ω	72V	0.6A
Total	150Ω	90V	0.6A

Exercise #21

Find the voltage and amperage for each resistor.

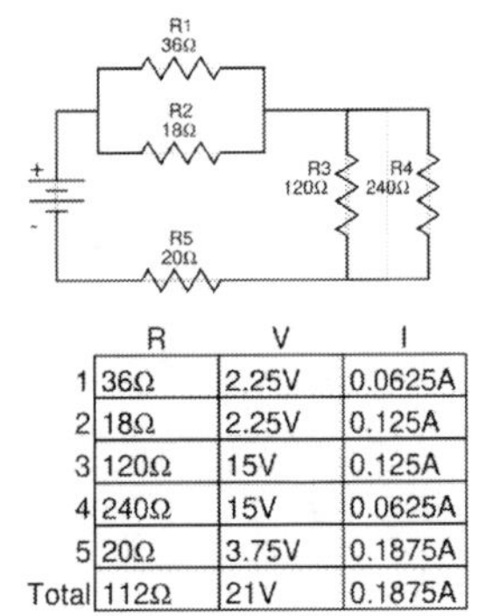

	R	V	I
1	36Ω	2.25V	0.0625A
2	18Ω	2.25V	0.125A
3	120Ω	15V	0.125A
4	240Ω	15V	0.0625A
5	20Ω	3.75V	0.1875A
Total	112Ω	21V	0.1875A

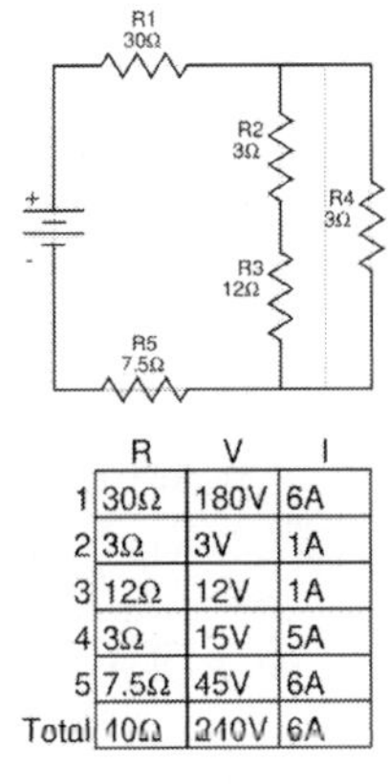

	R	V	I
1	30Ω	180V	6A
2	3Ω	3V	1A
3	12Ω	12V	1A
4	3Ω	15V	5A
5	7.5Ω	45V	6A
Total	10Ω	240V	6A

Exercise #22

Find the voltage and amperage for each resistor.

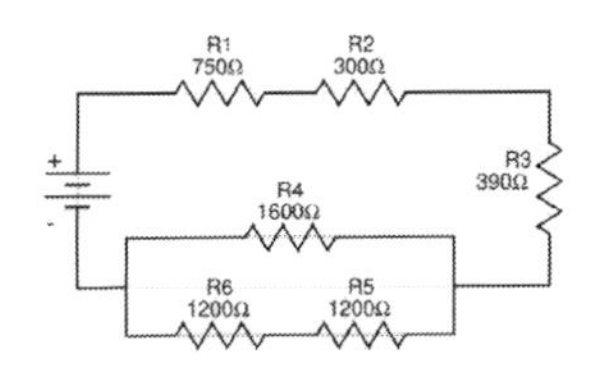

	R	V	I
1	750Ω	56.25V	0.075A
2	300Ω	22.5V	0.075A
3	390Ω	29.25V	0.075A
4	1600Ω	72V	0.045A
5	1200Ω	36V	0.03A
6	1200Ω	36V	0.03A
Total	2400Ω	180V	0.075A

Exercise #23

Find the voltage and amperage for each resistor.

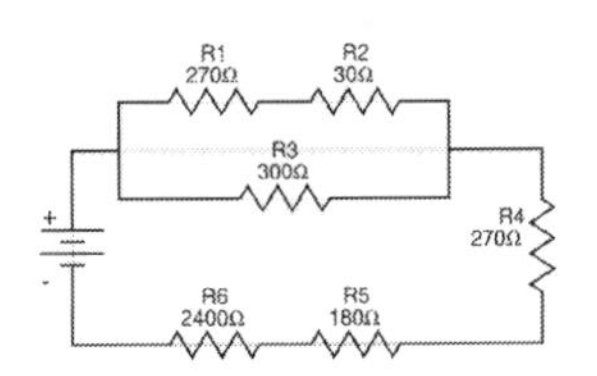

	R	V	I
1	270Ω	0.135V	0.0005A
2	30Ω	0.015V	0.0005A
3	300Ω	0.15V	0.0005A
4	270Ω	0.27V	0.001A
5	180Ω	0.18V	0.001A
6	2400Ω	2.4V	0.001A
Total	3000Ω	3V	0.001A

Exercise #24

Find the voltage and amperage for each resistor.

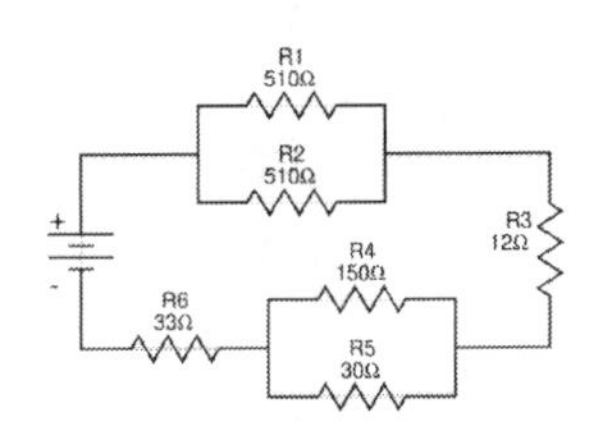

	R	V	I
1	510Ω	30.6V	0.06A
2	510Ω	30.6V	0.06A
3	12Ω	1.44V	0.12A
4	150Ω	3V	0.02A
5	30Ω	3V	0.1A
6	33Ω	3.96V	0.12A
Total	325Ω	39V	0.12A

Exercise #25

Find the voltage and amperage for each resistor.

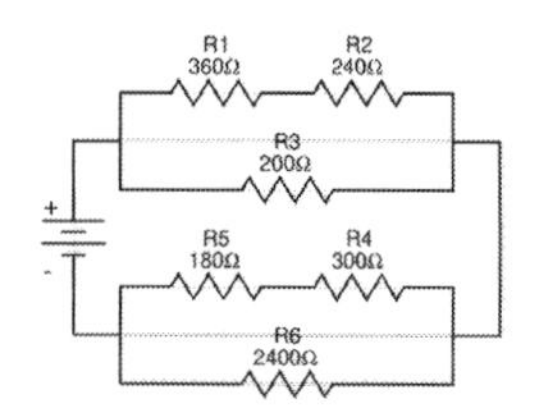

	R	V	I
1	360Ω	27V	0.075A
2	240Ω	18V	0.075A
3	200Ω	45V	0.225A
4	300Ω	75V	0.25A
5	180Ω	45V	0.25A
6	2400Ω	120V	0.05A
Total	550Ω	165V	0.3A

Exercise #26

Find the voltage and amperage for each resistor.

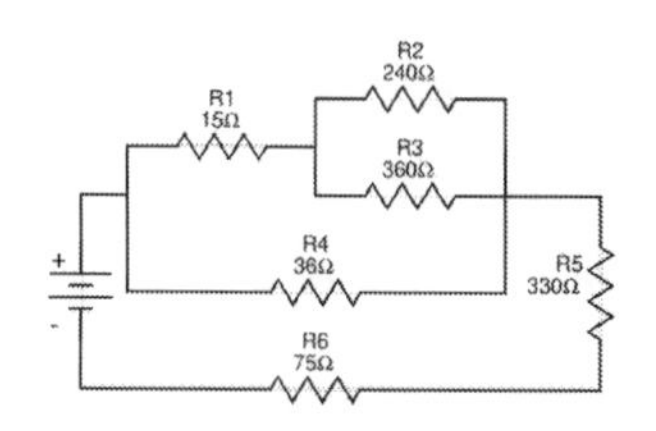

	R	V	I
1	15Ω	0.375V	0.025A
2	240Ω	3.6V	0.015A
3	360Ω	3.6V	0.01A
4	36Ω	0.9V	0.025A
5	330Ω	8.25V	0.025A
6	75Ω	1.875V	0.025A
Total	600Ω	15V	0.025A

Exercise #27

Find the voltage and amperage for each resistor.

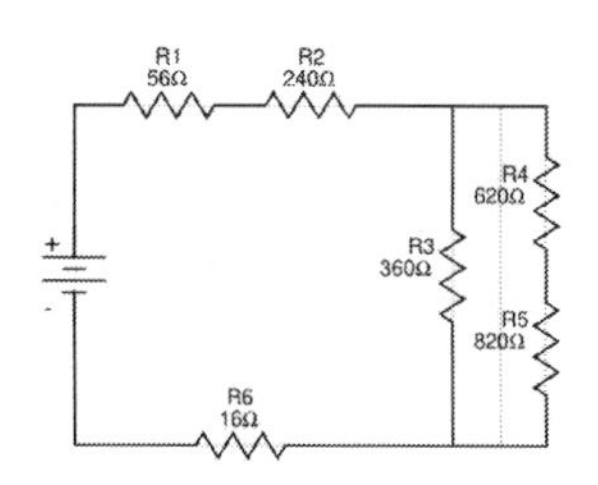

	R	V	I
1	56Ω	0.84V	0.015A
2	240Ω	3.6V	0.015A
3	360Ω	4.32V	0.012A
4	620Ω	1.86V	0.003A
5	820Ω	2.46V	0.003A
6	16Ω	0.24V	0.015A
Total	600Ω	9V	0.015A

Exercise #28

Find the voltage and amperage for each resistor.

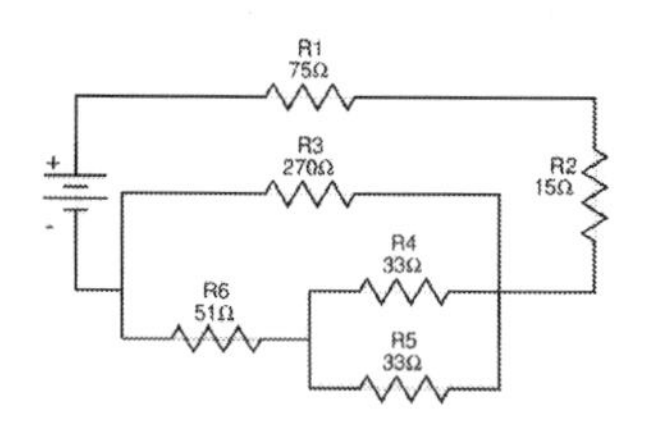

	R	V	I
1	75Ω	46.875V	0.625A
2	15Ω	9.375V	0.625A
3	270Ω	33.75V	0.125A
4	33Ω	8.25V	0.25A
5	33Ω	8.25V	0.25A
6	51Ω	25.5V	0.5A
Total	144Ω	90V	0.625A

Exercise #29

Find the voltage and amperage for each resistor.

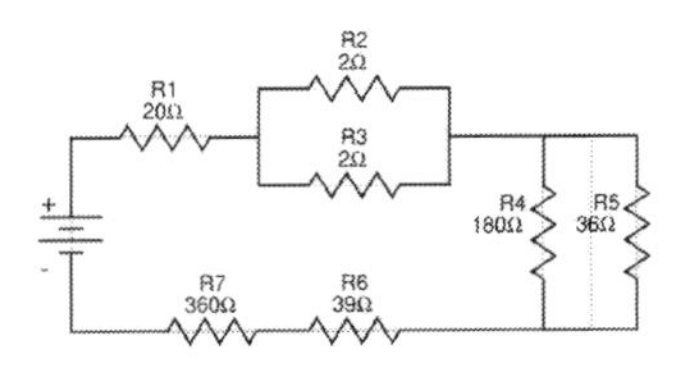

	R	V	I
1	20Ω	6V	0.3A
2	2Ω	0.3V	0.15A
3	2Ω	0.3V	0.15A
4	180Ω	9V	0.05A
5	36Ω	9V	0.25A
6	39Ω	11.7V	0.3A
7	360Ω	108V	0.3A
Total	450Ω	135V	0.3A

Exercise #30

Find the voltage and amperage for each resistor.

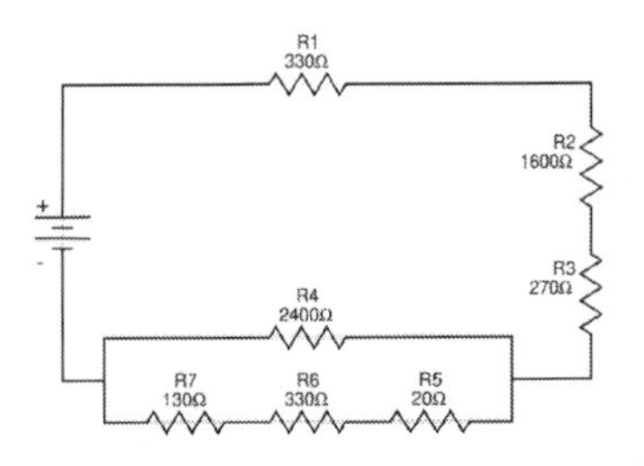

	R	V	I
1	330Ω	4.95V	0.015A
2	1600Ω	24V	0.015A
3	270Ω	4.05V	0.015A
4	2400Ω	6V	0.0025A
5	20Ω	0.25V	0.0125A
6	330Ω	4.125V	0.0125A
7	130Ω	1.625V	0.0125A
Total	2600Ω	39V	0.015A

Exercise #31

Find the voltage and amperage for each resistor.

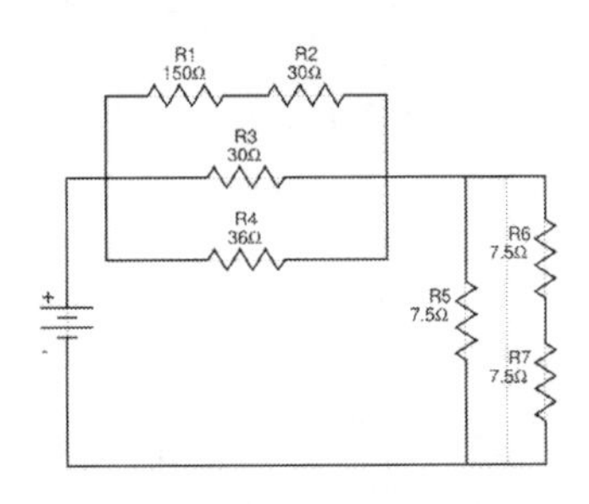

	R	V	I
1	150Ω	93.75V	0.625A
2	30Ω	18.75V	0.625A
3	30Ω	112.5V	3.75A
4	36Ω	112.5V	3.125A
5	7.5Ω	37.5V	5A
6	7.5Ω	18.75V	2.5A
7	7.5Ω	18.75V	2.5A
Total	20Ω	150V	7.5A

Exercise #32

Find the voltage and amperage for each resistor.

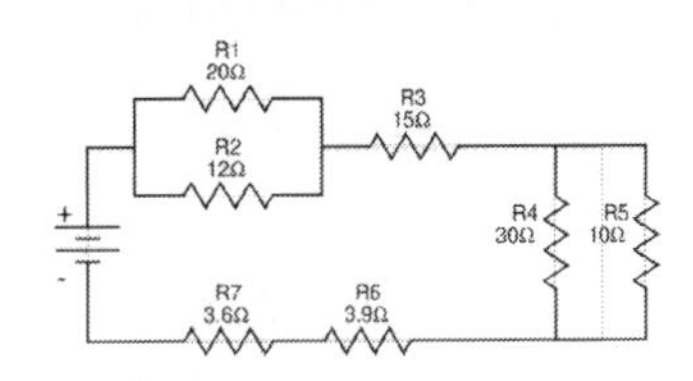

	R	V	I
1	20Ω	15V	0.75A
2	12Ω	15V	1.25A
3	15Ω	30V	2A
4	30Ω	15V	0.5A
5	10Ω	15V	1.5A
6	3.9Ω	7.8V	2A
7	3.6Ω	7.2V	2A
Total	37.5Ω	75V	2A

Exercise #33

Find the voltage and amperage for each resistor.

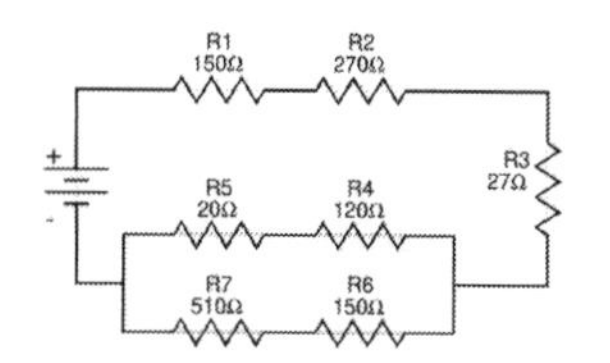

	R	V	I
1	150Ω	48V	0.32A
2	270Ω	86.4V	0.32A
3	27Ω	8.64V	0.32A
4	120Ω	31.68V	0.264A
5	20Ω	5.28V	0.264A
6	150Ω	8.4V	0.056A
7	510Ω	28.56V	0.056A
Total	562.5Ω	180V	0.32A

Exercise #34

Find the voltage and amperage for each resistor.

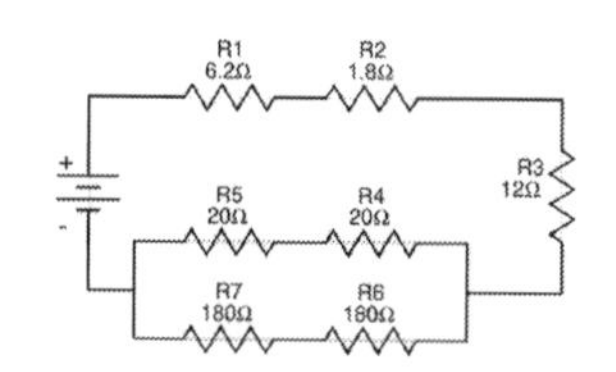

	R	V	I
1	6.2Ω	23.25V	3.75A
2	1.8Ω	6.75V	3.75A
3	12Ω	45V	3.75A
4	20Ω	67.5V	3.375A
5	20Ω	67.5V	3.375A
6	180Ω	67.5V	0.375A
7	180Ω	67.5V	0.375A
Total	56Ω	210V	3.75A

Exercise #35

Find the voltage and amperage for each resistor.

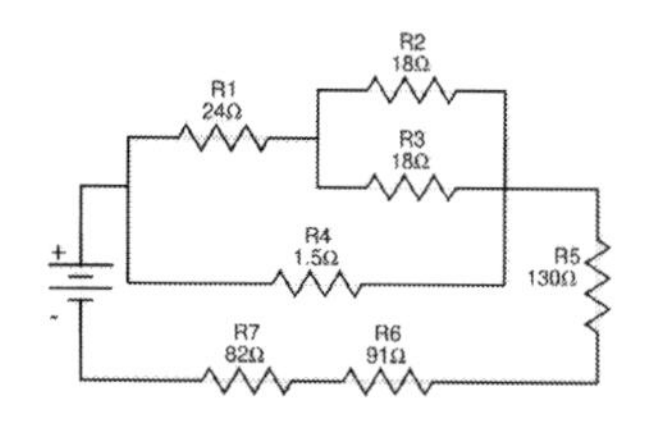

	R	V	I
1	24Ω	0.96V	0.04A
2	18Ω	0.36V	0.02A
3	18Ω	0.36V	0.02A
4	1.5Ω	0.06V	0.04A
5	130Ω	5.2V	0.04A
6	91Ω	3.64V	0.04A
7	82Ω	3.28V	0.04A
Total	337.5Ω	13.5V	0.04A

Exercise #36

Find the voltage and amperage for each resistor.

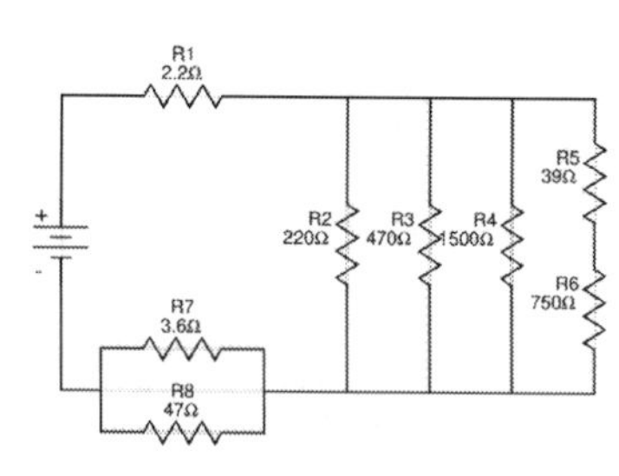

	R	V	I
1	2.2Ω	0.1084V	0.0493A
2	220Ω	5.7267V	0.026A
3	470Ω	5.7267V	0.0122A
4	1500Ω	5.7267V	0.0038A
5	39Ω	0.2831V	0.0073A
6	750Ω	5.4437V	0.0073A
7	3.6Ω	0.1648V	0.0458A
8	47Ω	0.1648V	0.0035A
Total	121.7256Ω	6V	0.0493A

Exercise #37

Find the voltage and amperage for each resistor.

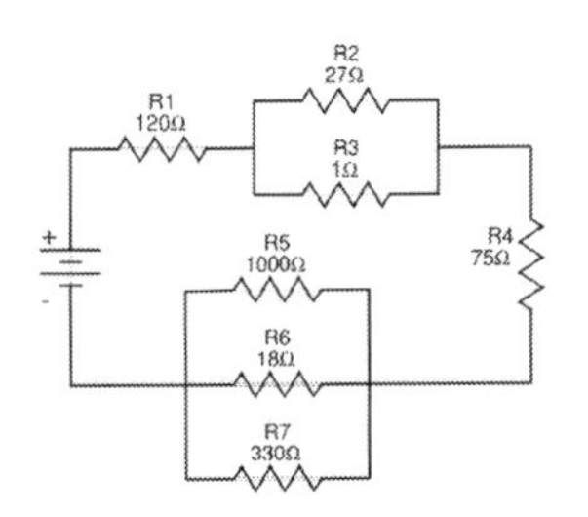

	R	V	I
1	120Ω	5.0765V	0.0423A
2	27Ω	0.0408V	0.0015A
3	1Ω	0.0408V	0.0408A
4	75Ω	3.1728V	0.0423A
5	1000Ω	0.71V	0.0007A
6	18Ω	0.71V	0.0394A
7	330Ω	0.71V	0.0022A
Total	212.7468Ω	9V	0.0423A

Exercise #38

Find the voltage and amperage for each resistor.

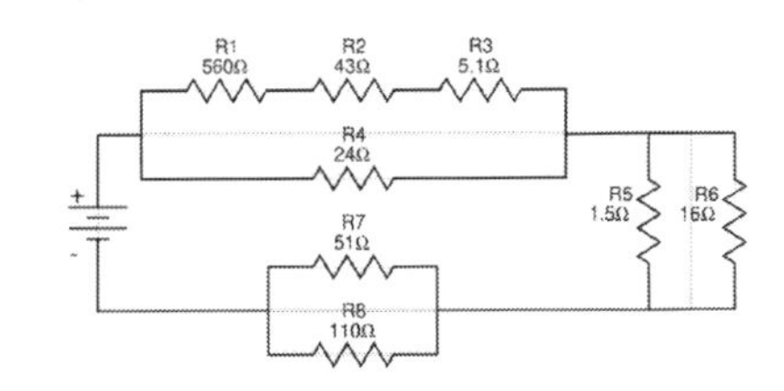

	R	V	I
1	560Ω	25.814V	0.0461A
2	43Ω	1.9821V	0.0461A
3	5.1Ω	0.2351V	0.0461A
4	24Ω	28.0312V	1.168A
5	1.5Ω	1.665V	1.11A
6	16Ω	1.665V	0.1041A
7	51Ω	42.3038V	0.8295A
8	110Ω	42.3038V	0.3846A
Total	59.3049Ω	72V	1.2141A

Exercise #39

Find the voltage and amperage for each resistor.

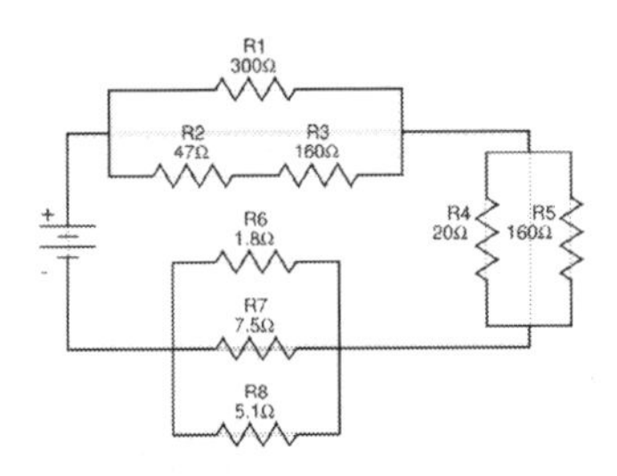

	R	V	I
1	300Ω	3.8982V	0.013A
2	47Ω	0.8851V	0.0188A
3	160Ω	3.0131V	0.0188A
4	20Ω	0.5658V	0.0283A
5	160Ω	0.5658V	0.0035A
6	1.8Ω	0.036V	0.02A
7	7.5Ω	0.036V	0.0048A
8	5.1Ω	0.036V	0.0071A
Total	141.393Ω	4.5V	0.0318A

Exercise #40

Find the voltage and amperage for each resistor.

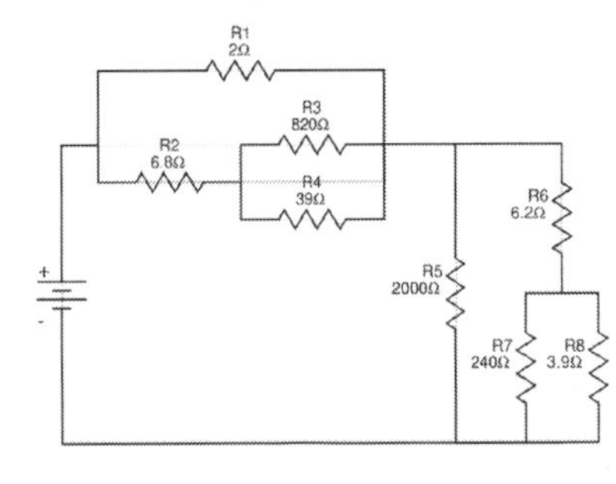

	R	V	I
1	2Ω	5.305V	2.6525A
2	6.8Ω	0.8193V	0.1205A
3	820Ω	4.4856V	0.0055A
4	39Ω	4.4856V	0.115A
5	2000Ω	27.695V	0.0138A
6	6.2Ω	17.1065V	2.7591A
7	240Ω	10.5885V	0.0441A
8	3.9Ω	10.5885V	2.715A
Total	11.9006Ω	33V	2.773A

Exercise #41

Find the voltage and amperage for each resistor.

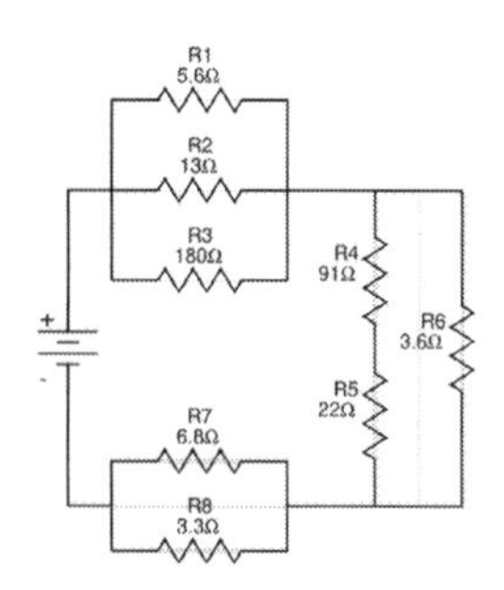

	R	V	I
1	5.6Ω	42.1558V	7.5278A
2	13Ω	42.1558V	3.2428A
3	180Ω	42.1558V	0.2342A
4	91Ω	30.9191V	0.3398A
5	22Ω	7.4749V	0.3398A
6	3.6Ω	38.394V	10.665A
7	6.8Ω	24.4502V	3.5956A
8	3.3Ω	24.4502V	7.4092A
Total	9.5413Ω	105V	11.0048A

Exercise #42

Find the voltage and amperage for each resistor.

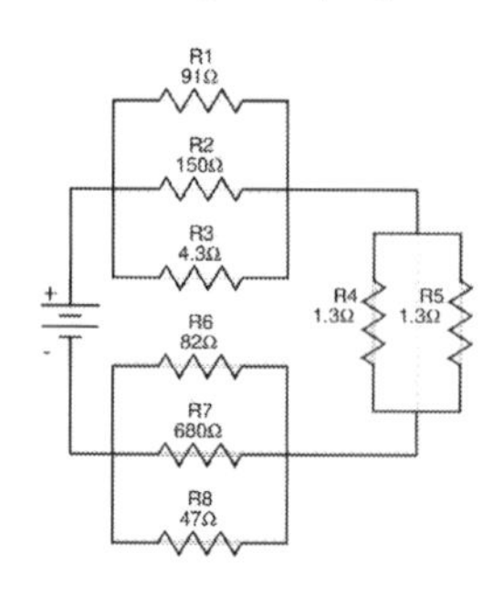

	R	V	I
1	91Ω	18.0215V	0.198A
2	150Ω	18.0215V	0.1201A
3	4.3Ω	18.0215V	4.191A
4	1.3Ω	2.931V	2.2546A
5	1.3Ω	2.931V	2.2546A
6	82Ω	129.0475V	1.5738A
7	680Ω	129.0475V	0.1898A
8	47Ω	129.0475V	2.7457A
Total	33.2652Ω	150V	4.5092A

Exercise #43

Find the voltage and amperage for each resistor.

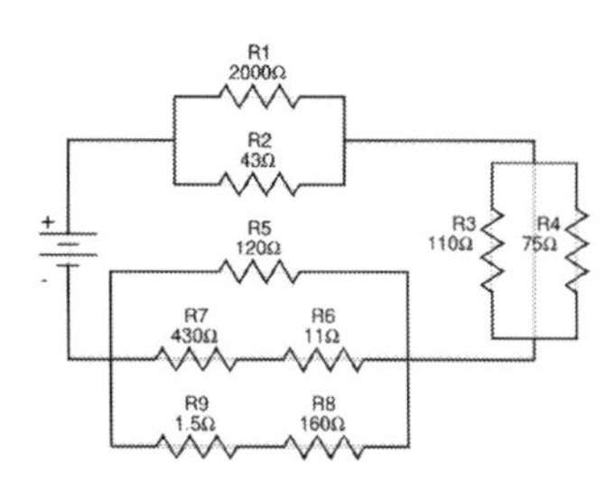

	R	V	I
1	2000Ω	2.1589V	0.0011A
2	43Ω	2.1589V	0.0502A
3	110Ω	2.2871V	0.0208A
4	75Ω	2.2871V	0.0305A
5	120Ω	3.054V	0.0255A
6	11Ω	0.0762V	0.0069A
7	430Ω	2.9779V	0.0069A
8	160Ω	3.0257V	0.0189A
9	1.5Ω	0.0284V	0.0189A
Total	146.2387Ω	7.5V	0.0513A

Exercise #44

Find the voltage and amperage for each resistor.

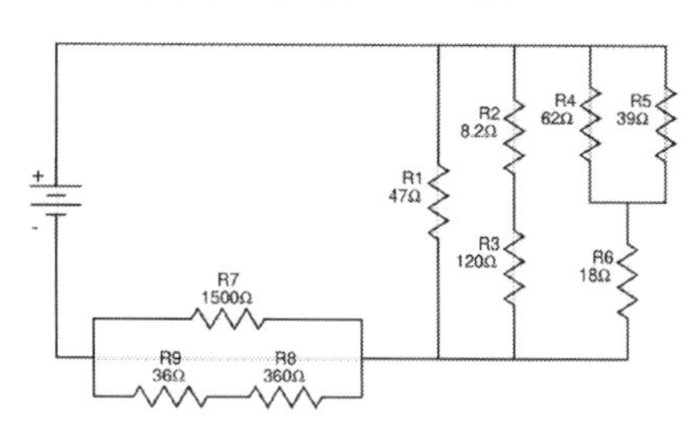

	R	V	I
1	47Ω	1.0239V	0.0218A
2	8.2Ω	0.0655V	0.008A
3	120Ω	0.9584V	0.008A
4	62Ω	0.5845V	0.0094A
5	39Ω	0.5845V	0.015A
6	18Ω	0.4394V	0.0244A
7	1500Ω	16.9761V	0.0113A
8	360Ω	15.4328V	0.0429A
9	36Ω	1.5433V	0.0429A
Total	332.1875Ω	18V	0.0542A

Exercise #45

Find the voltage and amperage for each resistor.

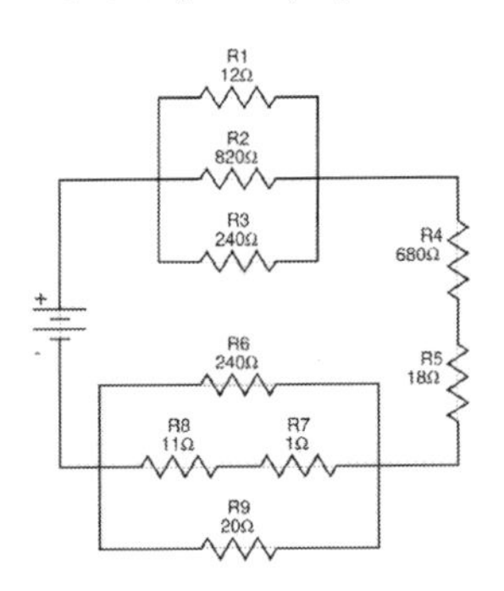

	R	V	I
1	12Ω	3.7753V	0.3146A
2	820Ω	3.7753V	0.0046A
3	240Ω	3.7753V	0.0157A
4	680Ω	227.7598V	0.3349A
5	18Ω	6.0289V	0.3349A
6	240Ω	2.4359V	0.0101A
7	1Ω	0.203V	0.203A
8	11Ω	2.2329V	0.203A
9	20Ω	2.4359V	0.1218A
Total	716.5442Ω	240V	0.3349A

Exercise #46

Find the voltage and amperage for each resistor.

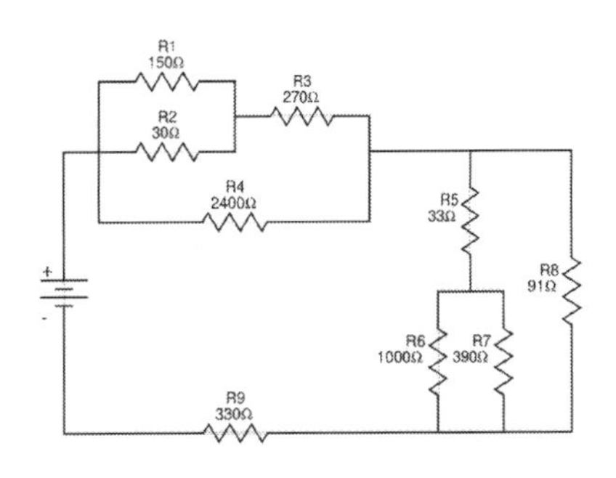

	R	V	I
1	150Ω	0.3021V	0.002A
2	30Ω	0.3021V	0.0101A
3	270Ω	3.2628V	0.0121A
4	2400Ω	3.5649V	0.0015A
5	33Ω	0.1007V	0.0031A
6	1000Ω	0.8564V	0.0009A
7	390Ω	0.8564V	0.0022A
8	91Ω	0.9571V	0.0105A
9	330Ω	4.478V	0.0136A
Total	663.2404Ω	9V	0.0136A

Exercise #47

Find the voltage and amperage for each resistor.

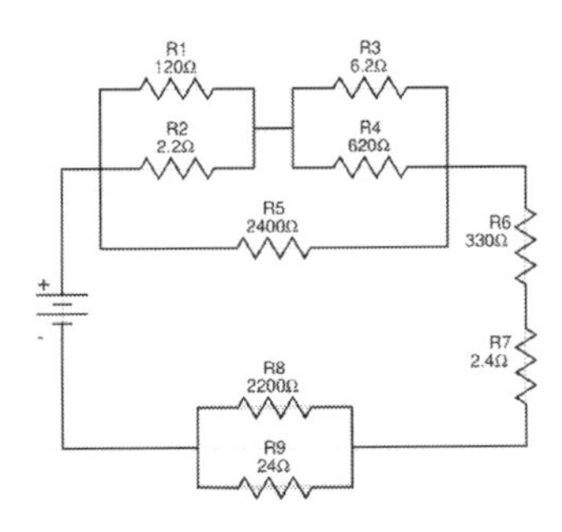

	R	V	I
1	120Ω	0.062V	0.0005A
2	2.2Ω	0.062V	0.0282A
3	6.2Ω	0.1763V	0.0284A
4	620Ω	0.1763V	0.0003A
5	2400Ω	0.2383V	0.0001A
6	330Ω	9.5085V	0.0288A
7	2.4Ω	0.0692V	0.0288A
8	2200Ω	0.6841V	0.0003A
9	24Ω	0.6841V	0.0285A
Total	364.4114Ω	10.5V	0.0288A

Exercise #48

Find the voltage and amperage for each resistor.

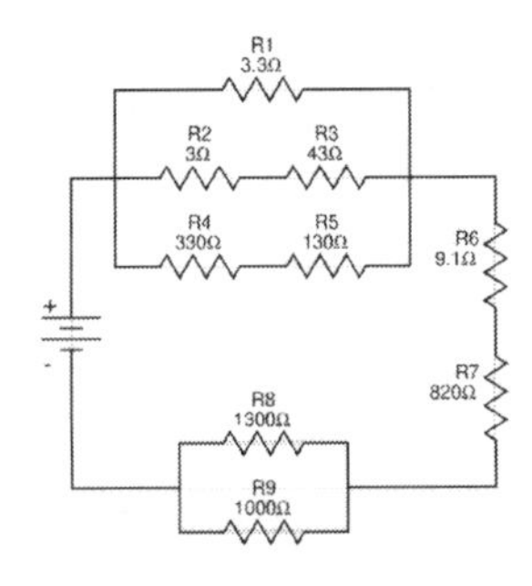

	R	V	I
1	3.3Ω	0.0657V	0.0199A
2	3Ω	0.0043V	0.0014A
3	43Ω	0.0614V	0.0014A
4	330Ω	0.0471V	0.0001A
5	130Ω	0.0186V	0.0001A
6	9.1Ω	0.1954V	0.0215A
7	820Ω	17.6044V	0.0215A
8	1300Ω	12.1345V	0.0093A
9	1000Ω	12.1345V	0.0121A
Total	1397.376Ω	30V	0.0215A

Exercise #49

Find the voltage and amperage for each resistor.

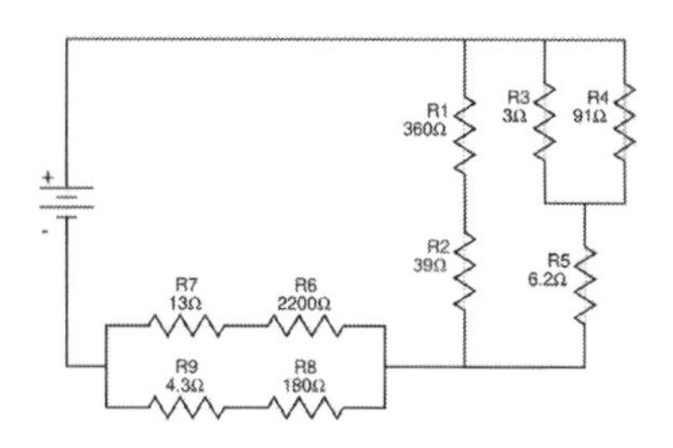

	R	V	I
1	360Ω	1.6149V	0.0045A
2	39Ω	0.1749V	0.0045A
3	3Ω	0.571V	0.1903A
4	91Ω	0.571V	0.0063A
5	6.2Ω	1.2189V	0.1966A
6	2200Ω	34.0092V	0.0155A
7	13Ω	0.201V	0.0155A
8	180Ω	33.412V	0.1856A
9	4.3Ω	0.7982V	0.1856A
Total	179.0325Ω	36V	0.2011A

Exercise #50

Find the voltage and amperage for each resistor.

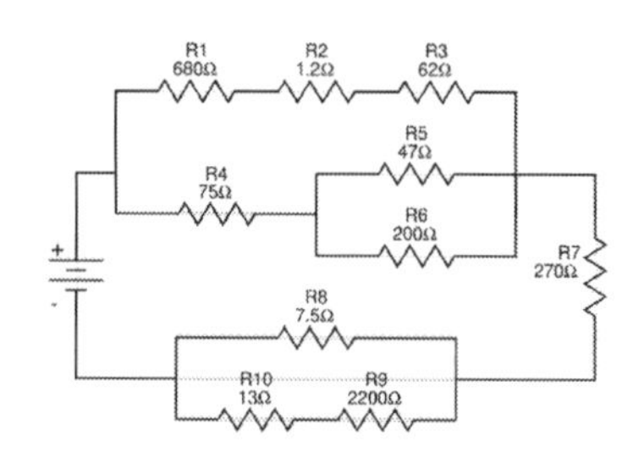

	R	V	I
1	680Ω	53.7841V	0.0791A
2	1.2Ω	0.0949V	0.0791A
3	62Ω	4.9038V	0.0791A
4	75Ω	38.9956V	0.5199A
5	47Ω	19.7872V	0.421A
6	200Ω	19.7872V	0.0989A
7	270Ω	161.7396V	0.599A
8	7.5Ω	4.4776V	0.597A
9	2200Ω	4.4513V	0.002A
10	13Ω	0.0263V	0.002A
Total	375.6038Ω	225V	0.599A

Exercise #51

Find the voltage and amperage for each resistor.

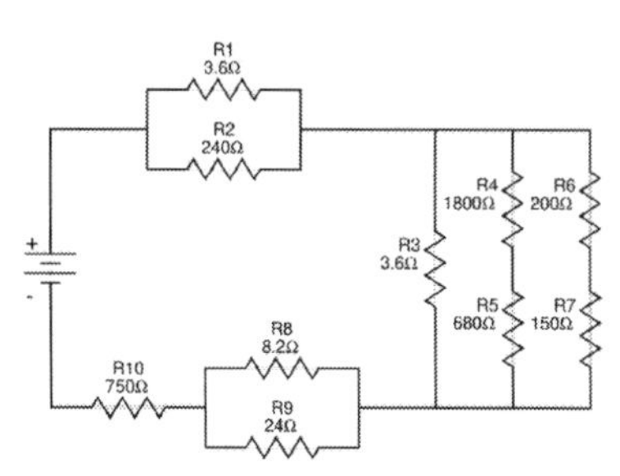

	R	V	I
1	3.6Ω	0.4182V	0.1162A
2	240Ω	0.4182V	0.0017A
3	3.6Ω	0.4196V	0.1166A
4	1800Ω	0.3045V	0.0002A
5	680Ω	0.115V	0.0002A
6	200Ω	0.2398V	0.0012A
7	150Ω	0.1798V	0.0012A
8	8.2Ω	0.7207V	0.0879A
9	24Ω	0.7207V	0.03A
10	750Ω	88.4414V	0.1179A
Total	763.2168Ω	90V	0.1179A

Exercise #52

Find the voltage and amperage for each resistor.

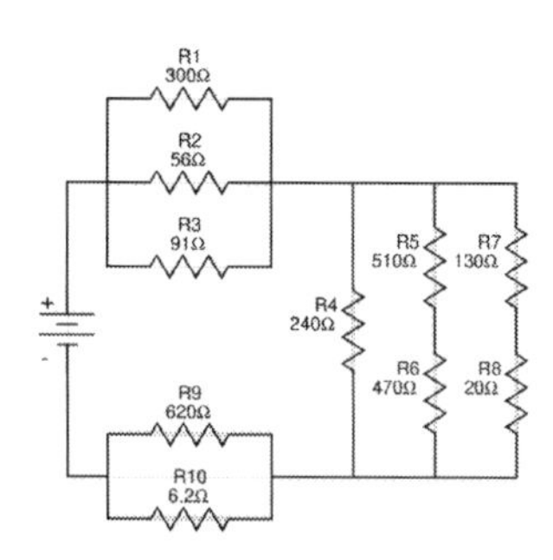

	R	V	I
1	300Ω	49.8435V	0.1661A
2	56Ω	49.8435V	0.8901A
3	91Ω	49.8435V	0.5477A
4	240Ω	135.3106V	0.5638A
5	510Ω	70.4167V	0.1381A
6	470Ω	64.8939V	0.1381A
7	130Ω	117.2692V	0.9021A
8	20Ω	18.0414V	0.9021A
9	620Ω	9.8459V	0.0159A
10	6.2Ω	9.8459V	1.5881A
Total	121.5759Ω	195V	1.6039A

Exercise #53

Find the voltage and amperage for each resistor.

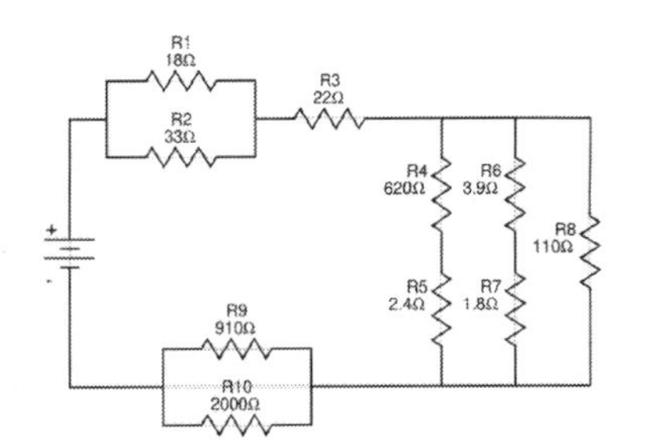

	R	V	I
1	18Ω	2.3664V	0.1315A
2	33Ω	2.3664V	0.0717A
3	22Ω	4.4699V	0.2032A
4	620Ω	1.0873V	0.0018A
5	2.4Ω	0.0042V	0.0018A
6	3.9Ω	0.7468V	0.1915A
7	1.8Ω	0.3447V	0.1915A
8	110Ω	1.0915V	0.0099A
9	910Ω	127.0722V	0.1396A
10	2000Ω	127.0722V	0.0635A
Total	664.449Ω	135V	0.2032A

Exercise #54

Find the voltage and amperage for each resistor.

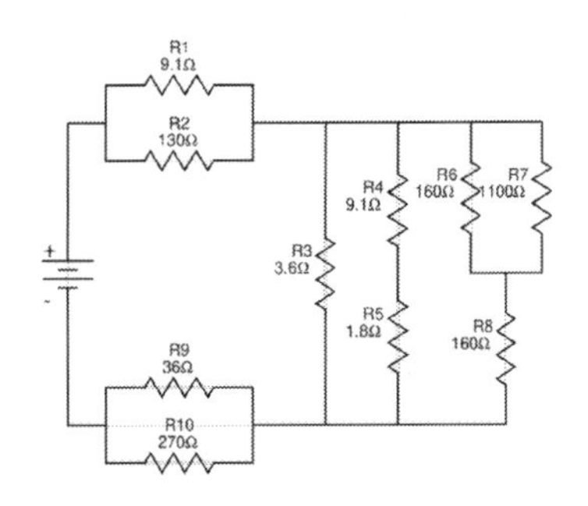

	R	V	I
1	9.1Ω	13.0685V	1.4361A
2	130Ω	13.0685V	0.1005A
3	3.6Ω	4.1212V	1.1448A
4	9.1Ω	3.4406V	0.3781A
5	1.8Ω	0.6806V	0.3781A
6	160Ω	1.9209V	0.012A
7	1100Ω	1.9209V	0.0017A
8	160Ω	2.2003V	0.0138A
9	36Ω	48.8103V	1.3558A
10	270Ω	48.8103V	0.1808A
Total	42.9514Ω	66V	1.5366A

Exercise #55

Find the voltage and amperage for each resistor.

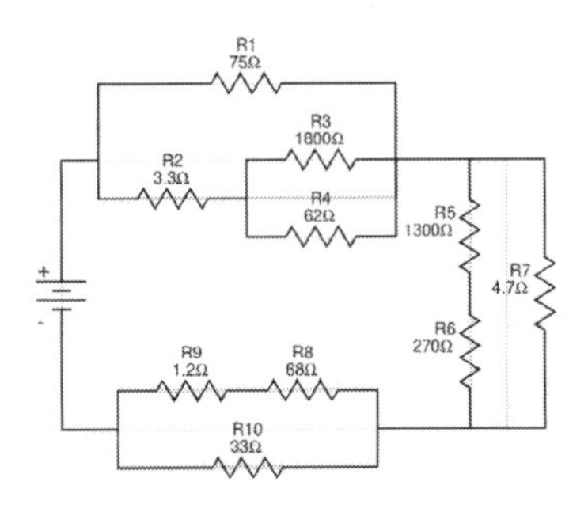

	R	V	I
1	75Ω	20.1358V	0.2685A
2	3.3Ω	1.0508V	0.3184A
3	1800Ω	19.085V	0.0106A
4	62Ω	19.085V	0.3078A
5	1300Ω	2.2772V	0.0018A
6	270Ω	0.473V	0.0018A
7	4.7Ω	2.7502V	0.5852A
8	68Ω	12.8866V	0.1895A
9	1.2Ω	0.2274V	0.1895A
10	33Ω	13.114V	0.3974A
Total	61.339Ω	36V	0.5869A

Exercise #56

Find the voltage and amperage for each resistor.

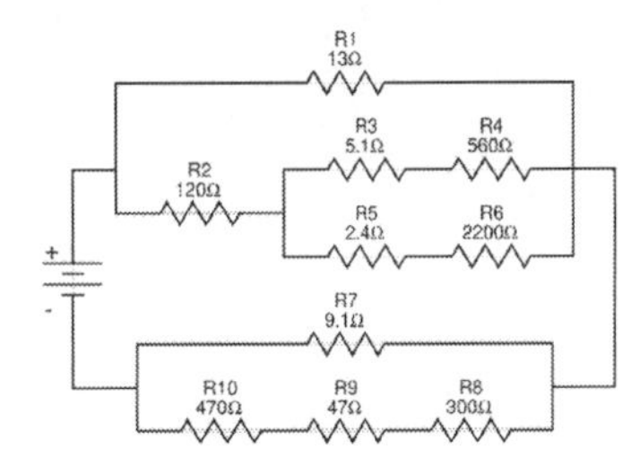

	R	V	I
1	13Ω	35.127V	2.7021A
2	120Ω	7.3989V	0.0617A
3	5.1Ω	0.2502V	0.0491A
4	560Ω	27.4779V	0.0491A
5	2.4Ω	0.0302V	0.0126A
6	2200Ω	27.6979V	0.0126A
7	9.1Ω	24.873V	2.7333A
8	300Ω	9.1333V	0.0304A
9	47Ω	1.4309V	0.0304A
10	470Ω	14.3088V	0.0304A
Total	21.7097Ω	60V	2.7637A

Exercise #57

Find the voltage and amperage for each resistor.

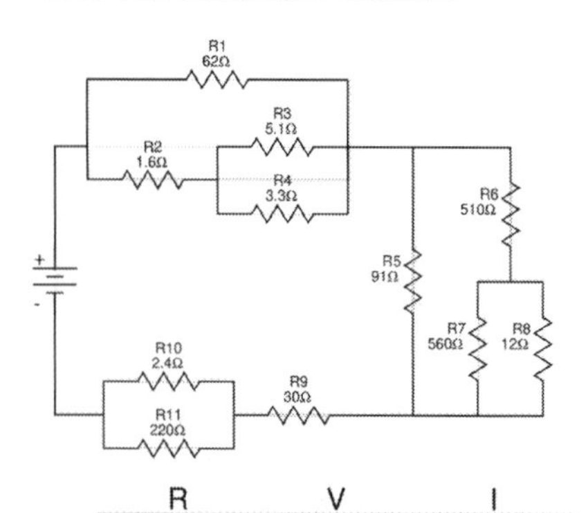

	R	V	I
1	62Ω	6.7652V	0.1091A
2	1.6Ω	3.0038V	1.8774A
3	5.1Ω	3.7614V	0.7375A
4	3.3Ω	3.7614V	1.1398A
5	91Ω	153.924V	1.6915A
6	510Ω	150.4581V	0.295A
7	560Ω	3.4659V	0.0062A
8	12Ω	3.4659V	0.2888A
9	30Ω	59.5946V	1.9865A
10	2.4Ω	4.7161V	1.9651A
11	220Ω	4.7161V	0.0214A
Total	113.2652Ω	225V	1.9865A

Exercise #58

Find the voltage and amperage for each resistor.

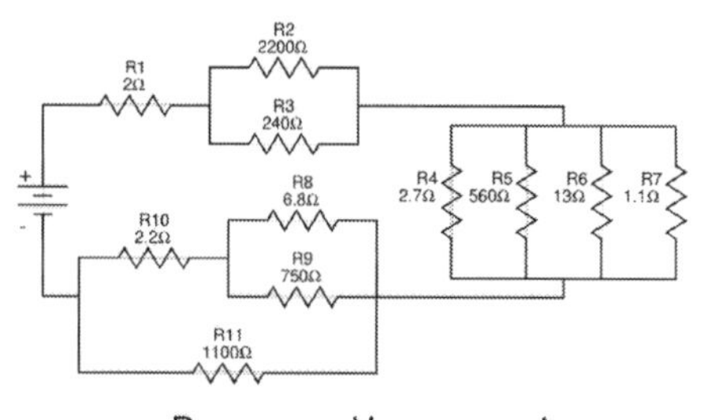

	R	V	I
1	2Ω	0.3158V	0.1579A
2	2200Ω	34.1679V	0.0155A
3	240Ω	34.1679V	0.1424A
4	2.7Ω	0.1163V	0.0431A
5	560Ω	0.1163V	0.0002A
6	13Ω	0.1163V	0.0089A
7	1.1Ω	0.1163V	0.1057A
8	6.8Ω	1.0555V	0.1552A
9	750Ω	1.0555V	0.0014A
10	2.2Ω	0.3446V	0.1566A
11	1100Ω	1.4V	0.0013A
Total	227.9966Ω	36V	0.1579A

Exercise #59

Find the voltage and amperage for each resistor.

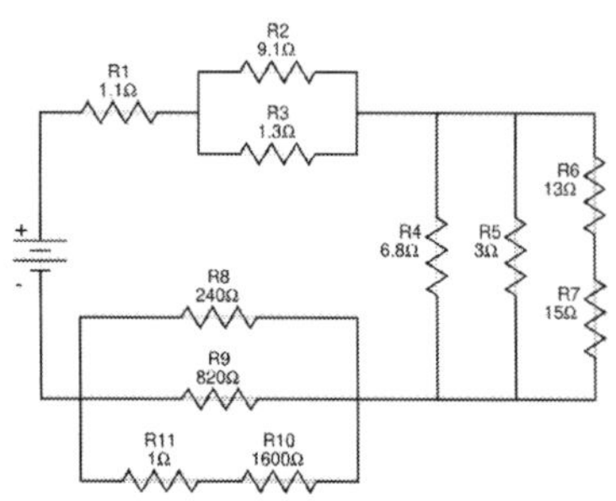

	R	V	I
1	1.1Ω	0.4838V	0.4398A
2	9.1Ω	0.5002V	0.055A
3	1.3Ω	0.5002V	0.3848A
4	6.8Ω	0.8521V	0.1253A
5	3Ω	0.8521V	0.284A
6	13Ω	0.3956V	0.0304A
7	15Ω	0.4565V	0.0304A
8	240Ω	73.1639V	0.3048A
9	820Ω	73.1639V	0.0892A
10	1600Ω	73.1182V	0.0457A
11	1Ω	0.0457V	0.0457A
Total	170.5426Ω	75V	0.4398A

Exercise #60

Find the voltage and amperage for each resistor.

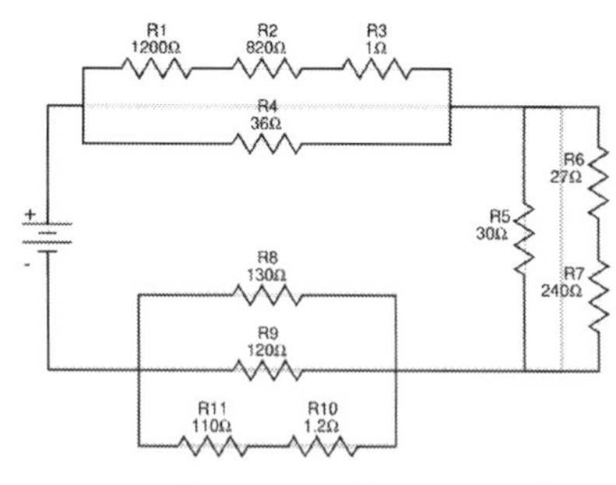

	R	V	I
1	1200Ω	0.3079V	0.0003A
2	820Ω	0.2104V	0.0003A
3	1Ω	0.0003V	0.0003A
4	36Ω	0.5186V	0.0144A
5	30Ω	0.3954V	0.0132A
6	27Ω	0.04V	0.0015A
7	240Ω	0.3554V	0.0015A
8	130Ω	0.586V	0.0045A
9	120Ω	0.586V	0.0049A
10	1.2Ω	0.0063V	0.0053A
11	110Ω	0.5797V	0.0053A
Total	102.3102Ω	1.5V	0.0147A

Exercise #61

Find the voltage and amperage for each resistor.

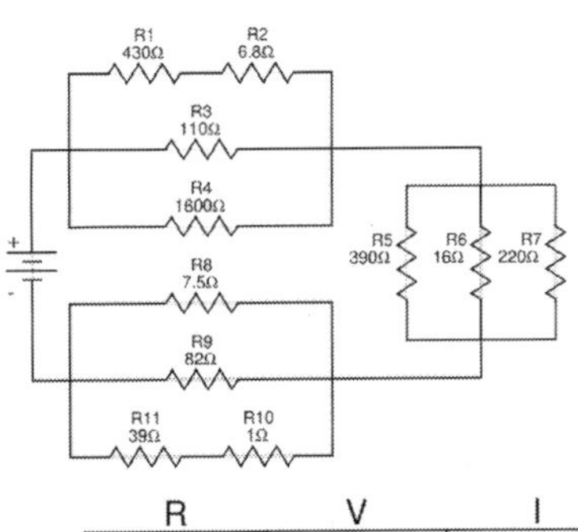

	R	V	I
1	430Ω	19.0096V	0.0442A
2	6.8Ω	0.3006V	0.0442A
3	110Ω	19.3102V	0.1755A
4	1600Ω	19.3102V	0.0121A
5	390Ω	3.3304V	0.0085A
6	16Ω	3.3304V	0.2081A
7	220Ω	3.3304V	0.0151A
8	7.5Ω	1.3594V	0.1813A
9	82Ω	1.3594V	0.0166A
10	1Ω	0.034V	0.034A
11	39Ω	1.3255V	0.034A
Total	103.5266Ω	24V	0.2318A

Exercise #62

Find the voltage and amperage for each resistor.

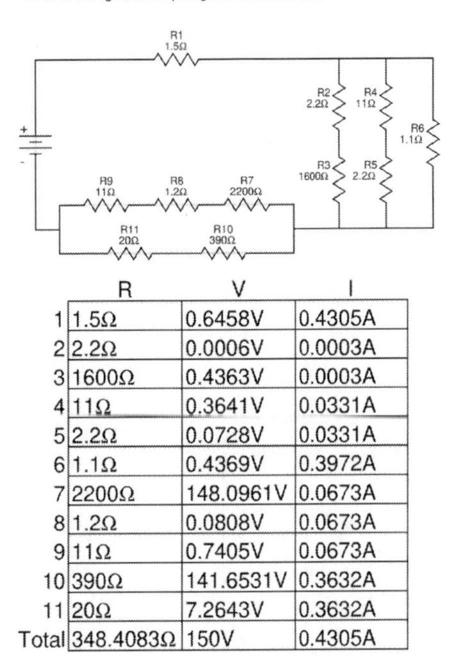

	R	V	I
1	1.5Ω	0.6458V	0.4305A
2	2.2Ω	0.0006V	0.0003A
3	1600Ω	0.4363V	0.0003A
4	11Ω	0.3641V	0.0331A
5	2.2Ω	0.0728V	0.0331A
6	1.1Ω	0.4369V	0.3972A
7	2200Ω	148.0961V	0.0673A
8	1.2Ω	0.0808V	0.0673A
9	11Ω	0.7405V	0.0673A
10	390Ω	141.6531V	0.3632A
11	20Ω	7.2643V	0.3632A
Total	348.4083Ω	150V	0.4305A

Exercise #63

Find the voltage and amperage for each resistor.

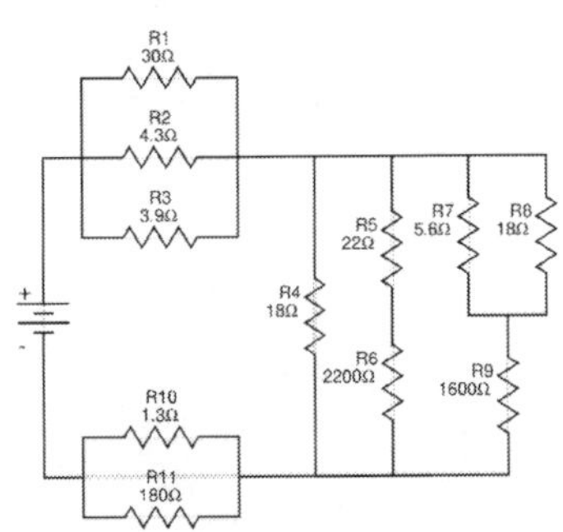

	R	V	I
1	30Ω	0.1376V	0.0046A
2	4.3Ω	0.1376V	0.032A
3	3.9Ω	0.1376V	0.0353A
4	18Ω	1.2696V	0.0705A
5	22Ω	0.0126V	0.0006A
6	2200Ω	1.257V	0.0006A
7	5.6Ω	0.0034V	0.0006A
8	18Ω	0.0034V	0.0002A
9	1600Ω	1.2662V	0.0008A
10	1.3Ω	0.0928V	0.0714A
11	180Ω	0.0928V	0.0005A
Total	20.8641Ω	1.5V	0.0719A

Exercise #64

Find the voltage and amperage for each resistor.

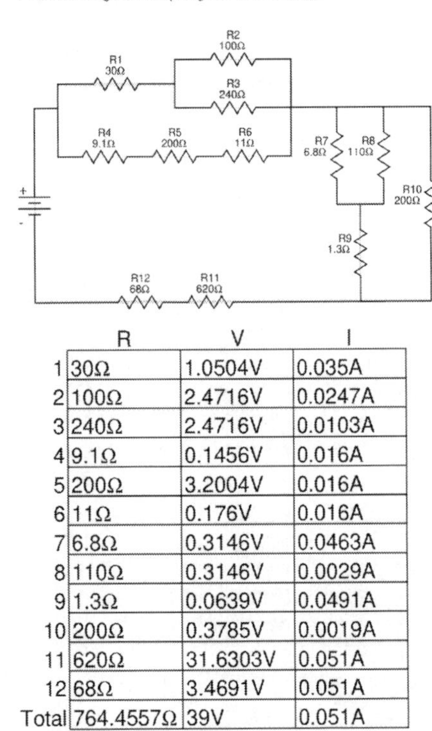

	R	V	I
1	30Ω	1.0504V	0.035A
2	100Ω	2.4716V	0.0247A
3	240Ω	2.4716V	0.0103A
4	9.1Ω	0.1456V	0.016A
5	200Ω	3.2004V	0.016A
6	11Ω	0.176V	0.016A
7	6.8Ω	0.3146V	0.0463A
8	110Ω	0.3146V	0.0029A
9	1.3Ω	0.0639V	0.0491A
10	200Ω	0.3785V	0.0019A
11	620Ω	31.6303V	0.051A
12	68Ω	3.4691V	0.051A
Total	764.4557Ω	39V	0.051A

Exercise #65

Find the voltage and amperage for each resistor.

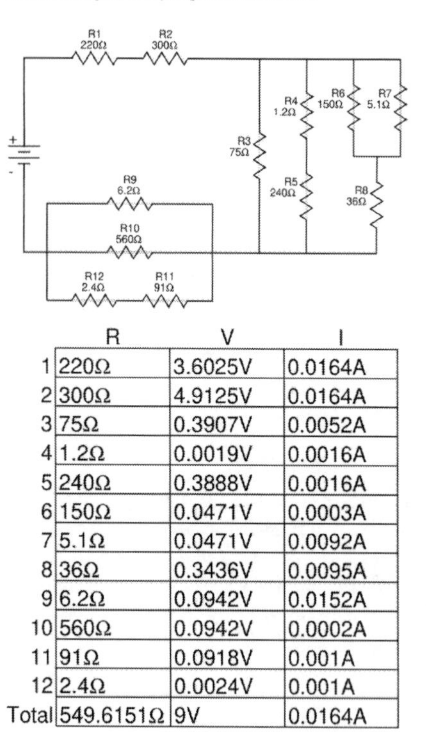

	R	V	I
1	220Ω	3.6025V	0.0164A
2	300Ω	4.9125V	0.0164A
3	75Ω	0.3907V	0.0052A
4	1.2Ω	0.0019V	0.0016A
5	240Ω	0.3888V	0.0016A
6	150Ω	0.0471V	0.0003A
7	5.1Ω	0.0471V	0.0092A
8	36Ω	0.3436V	0.0095A
9	6.2Ω	0.0942V	0.0152A
10	560Ω	0.0942V	0.0002A
11	91Ω	0.0918V	0.001A
12	2.4Ω	0.0024V	0.001A
Total	549.6151Ω	9V	0.0164A

Exercise #66

Find the voltage and amperage for each resistor.

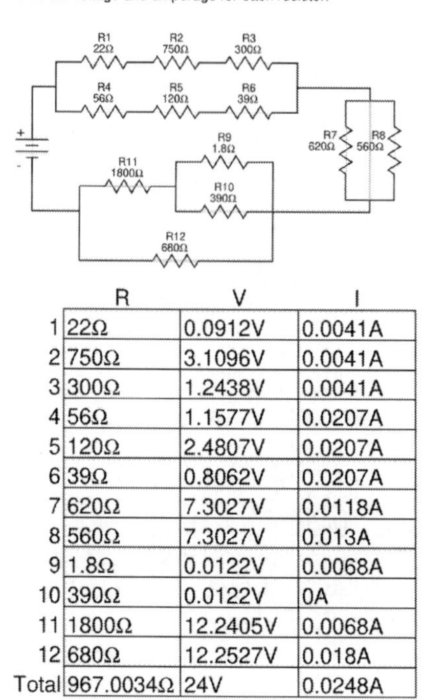

	R	V	I
1	22Ω	0.0912V	0.0041A
2	750Ω	3.1096V	0.0041A
3	300Ω	1.2438V	0.0041A
4	56Ω	1.1577V	0.0207A
5	120Ω	2.4807V	0.0207A
6	39Ω	0.8062V	0.0207A
7	620Ω	7.3027V	0.0118A
8	560Ω	7.3027V	0.013A
9	1.8Ω	0.0122V	0.0068A
10	390Ω	0.0122V	0A
11	1800Ω	12.2405V	0.0068A
12	680Ω	12.2527V	0.018A
Total	967.0034Ω	24V	0.0248A

Exercise #67

Find the voltage and amperage for each resistor.

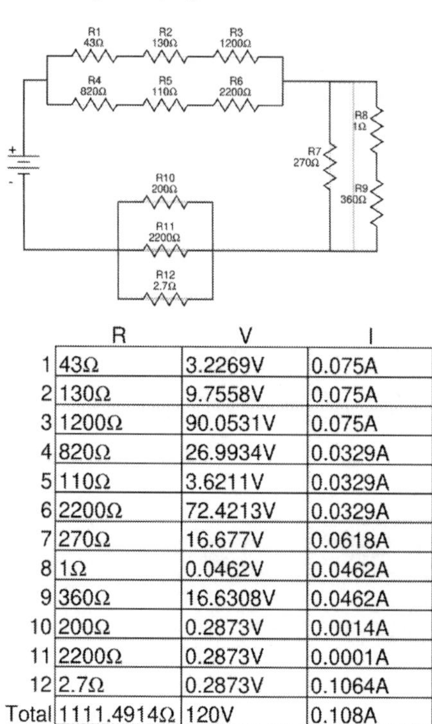

	R	V	I
1	43Ω	3.2269V	0.075A
2	130Ω	9.7558V	0.075A
3	1200Ω	90.0531V	0.075A
4	820Ω	26.9934V	0.0329A
5	110Ω	3.6211V	0.0329A
6	2200Ω	72.4213V	0.0329A
7	270Ω	16.677V	0.0618A
8	1Ω	0.0462V	0.0462A
9	360Ω	16.6308V	0.0462A
10	200Ω	0.2873V	0.0014A
11	2200Ω	0.2873V	0.0001A
12	2.7Ω	0.2873V	0.1064A
Total	1111.4914Ω	120V	0.108A

Exercise #68

Find the voltage and amperage for each resistor.

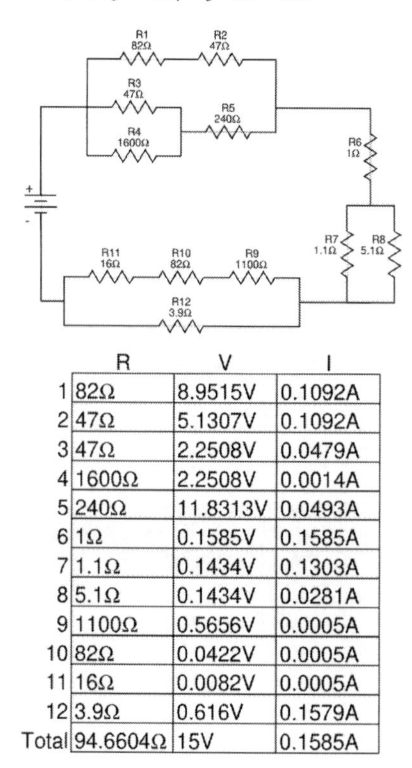

	R	V	I
1	82Ω	8.9515V	0.1092A
2	47Ω	5.1307V	0.1092A
3	47Ω	2.2508V	0.0479A
4	1600Ω	2.2508V	0.0014A
5	240Ω	11.8313V	0.0493A
6	1Ω	0.1585V	0.1585A
7	1.1Ω	0.1434V	0.1303A
8	5.1Ω	0.1434V	0.0281A
9	1100Ω	0.5656V	0.0005A
10	82Ω	0.0422V	0.0005A
11	16Ω	0.0082V	0.0005A
12	3.9Ω	0.616V	0.1579A
Total	94.6604Ω	15V	0.1585A

Exercise #69

Find the voltage and amperage for each resistor.

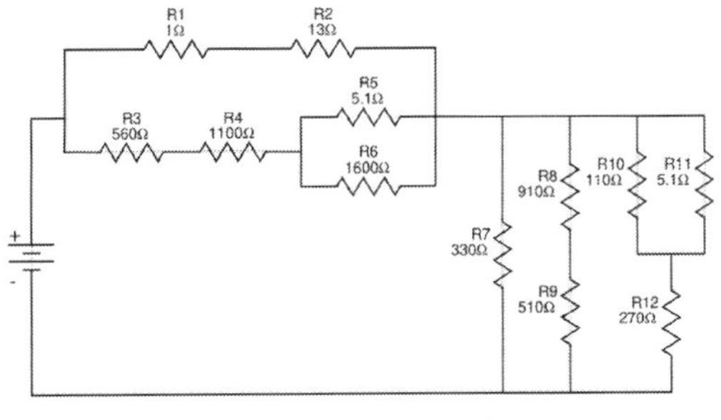

	R	V	I
1	1Ω	1.2933V	1.2933A
2	13Ω	16.8127V	1.2933A
3	560Ω	6.0894V	0.0109A
4	1100Ω	11.9613V	0.0109A
5	5.1Ω	0.0553V	0.0108A
6	1600Ω	0.0553V	0A
7	330Ω	176.894V	0.536A
8	910Ω	113.3616V	0.1246A
9	510Ω	63.5323V	0.1246A
10	110Ω	3.1367V	0.0285A
11	5.1Ω	3.1367V	0.615A
12	270Ω	173.7573V	0.6435A
Total	149.5214Ω	195V	1.3042A

Exercise #70

Find the voltage and amperage for each resistor.

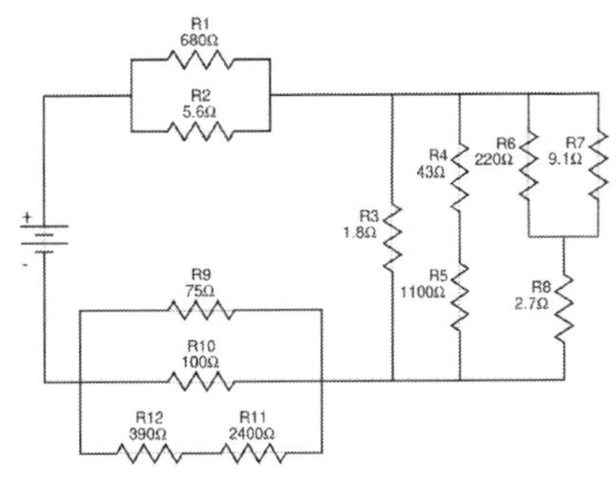

	R	V	I
1	680Ω	11.8257V	0.0174A
2	5.6Ω	11.8257V	2.1117A
3	1.8Ω	3.3068V	1.8371A
4	43Ω	0.1244V	0.0029A
5	1100Ω	3.1824V	0.0029A
6	220Ω	2.5263V	0.0115A
7	9.1Ω	2.5263V	0.2776A
8	2.7Ω	0.7806V	0.2891A
9	75Ω	89.8675V	1.1982A
10	100Ω	89.8675V	0.8987A
11	2400Ω	77.3054V	0.0322A
12	390Ω	12.5621V	0.0322A
Total	49.3162Ω	105V	2.1291A

Exercise #71

Find the voltage and amperage for each resistor.

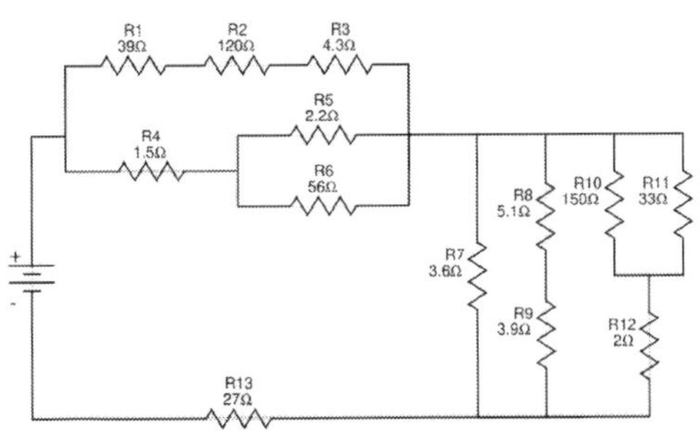

	R	V	I
1	39Ω	1.9264V	0.0494A
2	120Ω	5.9274V	0.0494A
3	4.3Ω	0.2124V	0.0494A
4	1.5Ω	3.3453V	2.2302A
5	2.2Ω	4.7209V	2.1459A
6	56Ω	4.7209V	0.0843A
7	3.6Ω	5.3851V	1.4959A
8	5.1Ω	3.0516V	0.5983A
9	3.9Ω	2.3335V	0.5983A
10	150Ω	5.0143V	0.0334A
11	33Ω	5.0143V	0.1519A
12	2Ω	0.3708V	0.1854A
13	27Ω	61.5487V	2.2796A
Total	32.9008Ω	75V	2.2796A

Exercise #72

Find the voltage and amperage for each resistor.

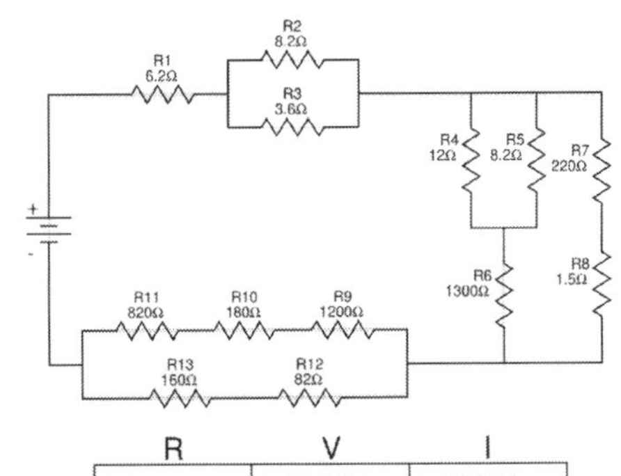

	R	V	I
1	6.2Ω	0.5364V	0.0865A
2	8.2Ω	0.2165V	0.0264A
3	3.6Ω	0.2165V	0.0601A
4	12Ω	0.0612V	0.0051A
5	8.2Ω	0.0612V	0.0075A
6	1300Ω	16.3225V	0.0126A
7	220Ω	16.2727V	0.074A
8	1.5Ω	0.111V	0.074A
9	1200Ω	10.2892V	0.0086A
10	180Ω	1.5434V	0.0086A
11	820Ω	7.0309V	0.0086A
12	82Ω	6.3918V	0.0779A
13	160Ω	12.4717V	0.0779A
Total	416.0766Ω	36V	0.0865A

Exercise #73

Find the voltage and amperage for each resistor.

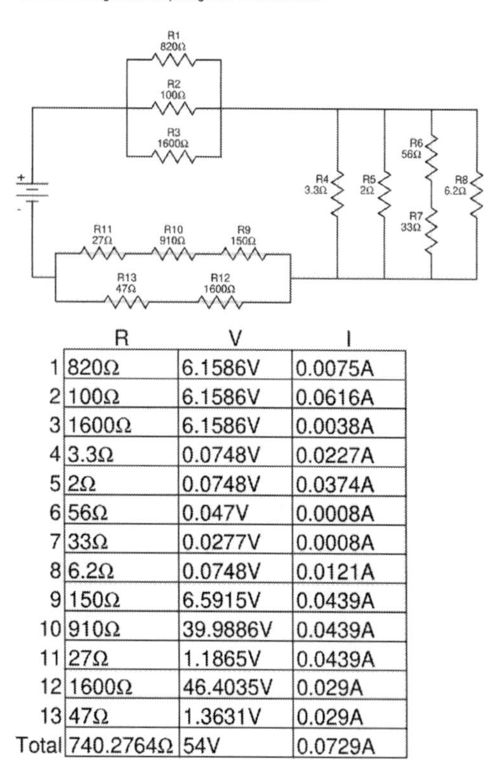

	R	V	I
1	820Ω	6.1586V	0.0075A
2	100Ω	6.1586V	0.0616A
3	1600Ω	6.1586V	0.0038A
4	3.3Ω	0.0748V	0.0227A
5	2Ω	0.0748V	0.0374A
6	56Ω	0.047V	0.0008A
7	33Ω	0.0277V	0.0008A
8	6.2Ω	0.0748V	0.0121A
9	150Ω	6.5915V	0.0439A
10	910Ω	39.9886V	0.0439A
11	27Ω	1.1865V	0.0439A
12	1600Ω	46.4035V	0.029A
13	47Ω	1.3631V	0.029A
Total	740.2764Ω	54V	0.0729A

Exercise #74

Find the voltage and amperage for each resistor.

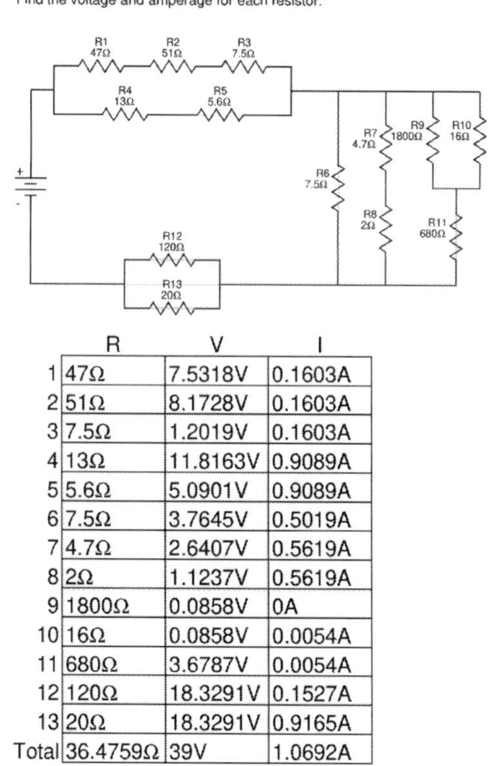

	R	V	I
1	47Ω	7.5318V	0.1603A
2	51Ω	8.1728V	0.1603A
3	7.5Ω	1.2019V	0.1603A
4	13Ω	11.8163V	0.9089A
5	5.6Ω	5.0901V	0.9089A
6	7.5Ω	3.7645V	0.5019A
7	4.7Ω	2.6407V	0.5619A
8	2Ω	1.1237V	0.5619A
9	1800Ω	0.0858V	0A
10	16Ω	0.0858V	0.0054A
11	680Ω	3.6787V	0.0054A
12	120Ω	18.3291V	0.1527A
13	20Ω	18.3291V	0.9165A
Total	36.4759Ω	39V	1.0692A

Exercise #75

Find the voltage and amperage for each resistor.

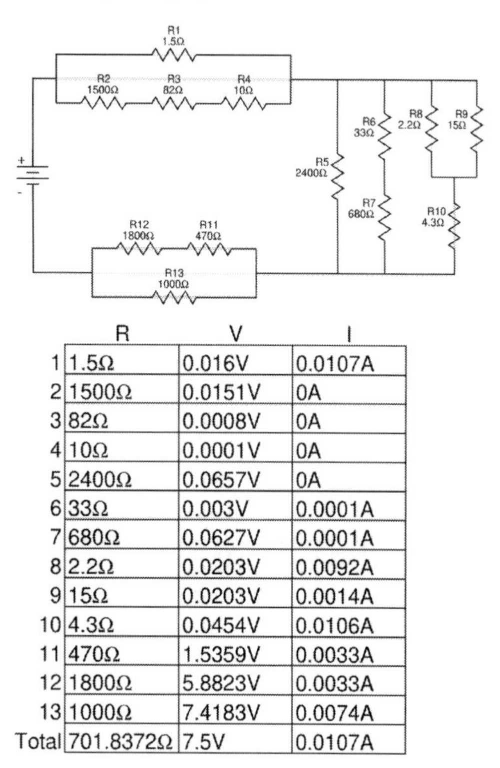

	R	V	I
1	1.5Ω	0.016V	0.0107A
2	1500Ω	0.0151V	0A
3	82Ω	0.0008V	0A
4	10Ω	0.0001V	0A
5	2400Ω	0.0657V	0A
6	33Ω	0.003V	0.0001A
7	680Ω	0.0627V	0.0001A
8	2.2Ω	0.0203V	0.0092A
9	15Ω	0.0203V	0.0014A
10	4.3Ω	0.0454V	0.0106A
11	470Ω	1.5359V	0.0033A
12	1800Ω	5.8823V	0.0033A
13	1000Ω	7.4183V	0.0074A
Total	701.8372Ω	7.5V	0.0107A

Exercise #76

Find the voltage and amperage for each resistor.

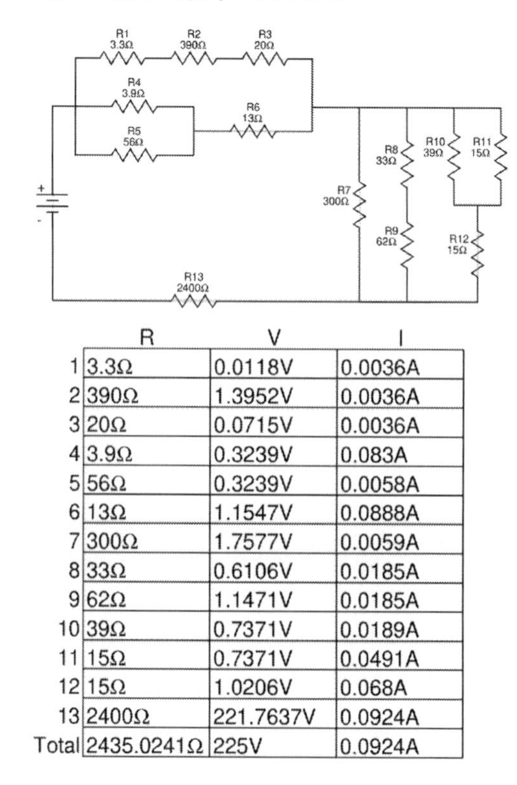

	R	V	I
1	3.3Ω	0.0118V	0.0036A
2	390Ω	1.3952V	0.0036A
3	20Ω	0.0715V	0.0036A
4	3.9Ω	0.3239V	0.083A
5	56Ω	0.3239V	0.0058A
6	13Ω	1.1547V	0.0888A
7	300Ω	1.7577V	0.0059A
8	33Ω	0.6106V	0.0185A
9	62Ω	1.1471V	0.0185A
10	39Ω	0.7371V	0.0189A
11	15Ω	0.7371V	0.0491A
12	15Ω	1.0206V	0.068A
13	2400Ω	221.7637V	0.0924A
Total	2435.0241Ω	225V	0.0924A

Exercise #77

Find the voltage and amperage for each resistor.

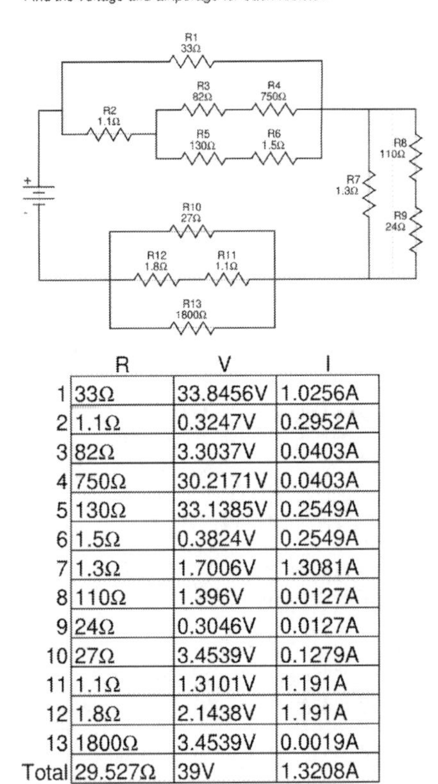

	R	V	I
1	33Ω	33.8456V	1.0256A
2	1.1Ω	0.3247V	0.2952A
3	82Ω	3.3037V	0.0403A
4	750Ω	30.2171V	0.0403A
5	130Ω	33.1385V	0.2549A
6	1.5Ω	0.3824V	0.2549A
7	1.3Ω	1.7006V	1.3081A
8	110Ω	1.396V	0.0127A
9	24Ω	0.3046V	0.0127A
10	27Ω	3.4539V	0.1279A
11	1.1Ω	1.3101V	1.191A
12	1.8Ω	2.1438V	1.191A
13	1800Ω	3.4539V	0.0019A
Total	29.527Ω	39V	1.3208A

Exercise #78

Find the voltage and amperage for each resistor.

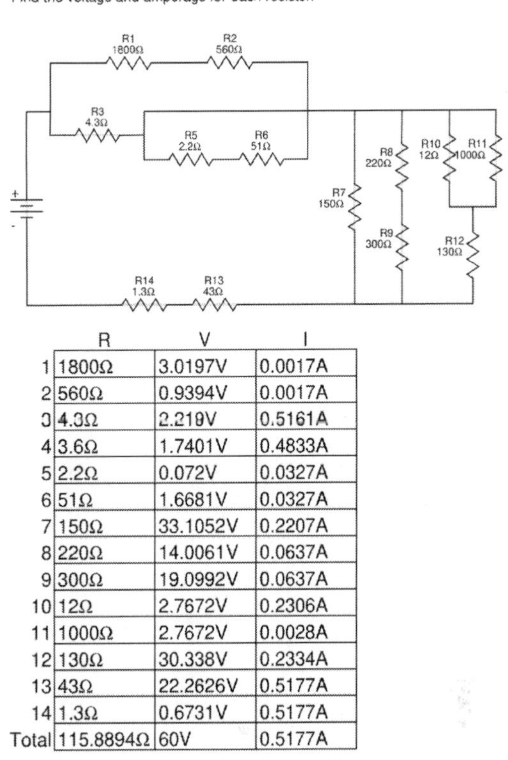

	R	V	I
1	1800Ω	3.0197V	0.0017A
2	560Ω	0.9394V	0.0017A
3	4.3Ω	2.219V	0.5161A
4	3.6Ω	1.7401V	0.4833A
5	2.2Ω	0.072V	0.0327A
6	51Ω	1.6681V	0.0327A
7	150Ω	33.1052V	0.2207A
8	220Ω	14.0061V	0.0637A
9	300Ω	19.0992V	0.0637A
10	12Ω	2.7672V	0.2306A
11	1000Ω	2.7672V	0.0028A
12	130Ω	30.338V	0.2334A
13	43Ω	22.2626V	0.5177A
14	1.3Ω	0.6731V	0.5177A
Total	115.8894Ω	60V	0.5177A

Exercise #79

Find the voltage and amperage for each resistor.

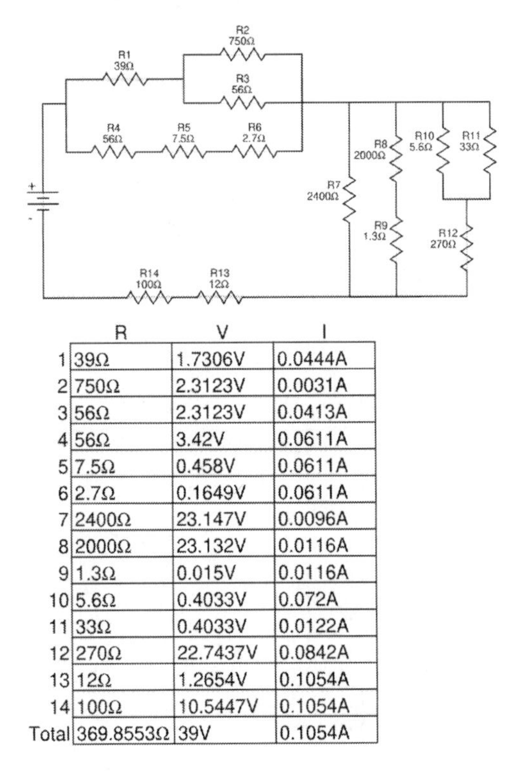

	R	V	I
1	39Ω	1.7306V	0.0444A
2	750Ω	2.3123V	0.0031A
3	56Ω	2.3123V	0.0413A
4	56Ω	3.42V	0.0611A
5	7.5Ω	0.458V	0.0611A
6	2.7Ω	0.1649V	0.0611A
7	2400Ω	23.147V	0.0096A
8	2000Ω	23.132V	0.0116A
9	1.3Ω	0.015V	0.0116A
10	5.6Ω	0.4033V	0.072A
11	33Ω	0.4033V	0.0122A
12	270Ω	22.7437V	0.0842A
13	12Ω	1.2654V	0.1054A
14	100Ω	10.5447V	0.1054A
Total	369.8553Ω	39V	0.1054A

Exercise #80

Find the voltage and amperage for each resistor.

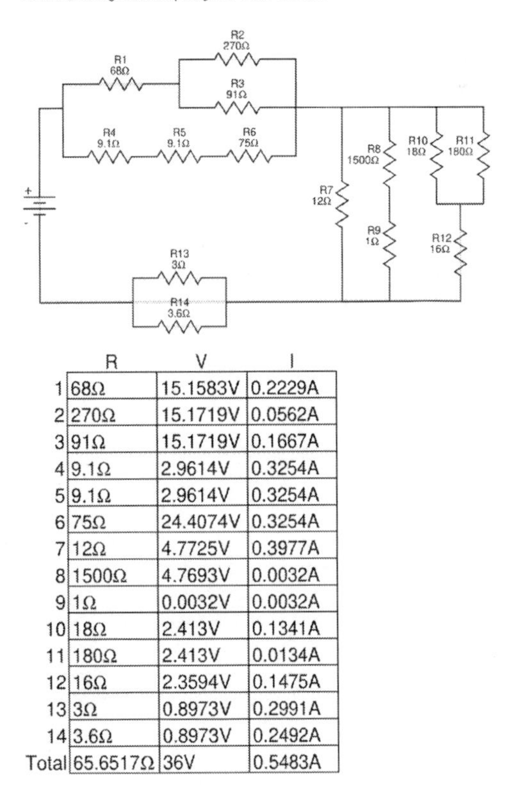

	R	V	I
1	68Ω	15.1583V	0.2229A
2	270Ω	15.1719V	0.0562A
3	91Ω	15.1719V	0.1667A
4	9.1Ω	2.9614V	0.3254A
5	9.1Ω	2.9614V	0.3254A
6	75Ω	24.4074V	0.3254A
7	12Ω	4.7725V	0.3977A
8	1500Ω	4.7693V	0.0032A
9	1Ω	0.0032V	0.0032A
10	18Ω	2.413V	0.1341A
11	180Ω	2.413V	0.0134A
12	16Ω	2.3594V	0.1475A
13	3Ω	0.8973V	0.2991A
14	3.6Ω	0.8973V	0.2492A
Total	65.6517Ω	36V	0.5483A

Exercise #81

Find the voltage and amperage for each resistor.

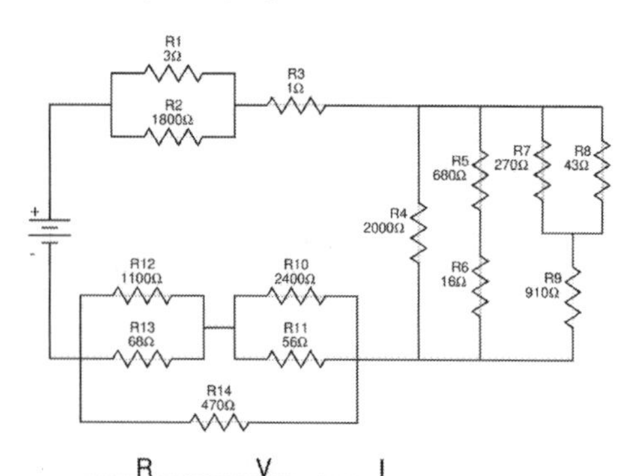

	R	V	I
1	3Ω	0.2698V	0.0899A
2	1800Ω	0.2698V	0.0001A
3	1Ω	0.0901V	0.0901A
4	2000Ω	30.1V	0.0151A
5	680Ω	29.4081V	0.0432A
6	16Ω	0.692V	0.0432A
7	270Ω	1.1789V	0.0044A
8	43Ω	1.1789V	0.0274A
9	910Ω	28.9212V	0.0318A
10	2400Ω	3.935V	0.0016A
11	56Ω	3.935V	0.0703A
12	1100Ω	4.6051V	0.0042A
13	68Ω	4.6051V	0.0677A
14	470Ω	8.5401V	0.0182A
Total	432.955Ω	39V	0.0901A

Exercise #82

Find the voltage and amperage for each resistor.

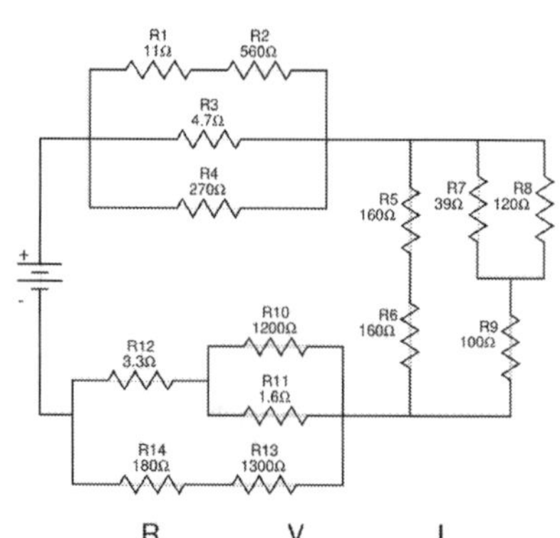

	R	V	I
1	11Ω	0.0208V	0.0019A
2	560Ω	1.0614V	0.0019A
3	4.7Ω	1.0822V	0.2303A
4	270Ω	1.0822V	0.004A
5	160Ω	10.8824V	0.068A
6	160Ω	10.8824V	0.068A
7	39Ω	4.9494V	0.1269A
8	120Ω	4.9494V	0.0412A
9	100Ω	16.8154V	0.1682A
10	1200Ω	0.3761V	0.0003A
11	1.6Ω	0.3761V	0.2351A
12	3.3Ω	0.7768V	0.2354A
13	1300Ω	1.0127V	0.0008A
14	180Ω	0.1402V	0.0008A
Total	101.6221Ω	24V	0.2362A

Exercise #83

Find the voltage and amperage for each resistor.

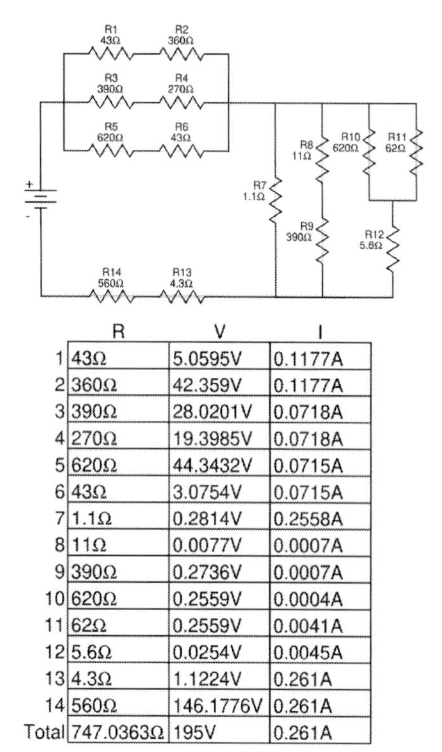

	R	V	I
1	43Ω	5.0595V	0.1177A
2	360Ω	42.359V	0.1177A
3	390Ω	28.0201V	0.0718A
4	270Ω	19.3985V	0.0718A
5	620Ω	44.3432V	0.0715A
6	43Ω	3.0754V	0.0715A
7	1.1Ω	0.2814V	0.2558A
8	11Ω	0.0077V	0.0007A
9	390Ω	0.2736V	0.0007A
10	620Ω	0.2559V	0.0004A
11	62Ω	0.2559V	0.0041A
12	5.6Ω	0.0254V	0.0045A
13	4.3Ω	1.1224V	0.261A
14	560Ω	146.1776V	0.261A
Total	747.0363Ω	195V	0.261A

Exercise #84

Find the voltage and amperage for each resistor.

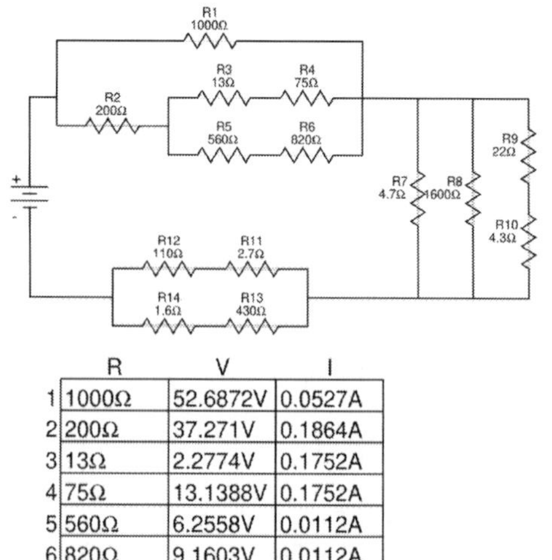

	R	V	I
1	1000Ω	52.6872V	0.0527A
2	200Ω	37.271V	0.1864A
3	13Ω	2.2774V	0.1752A
4	75Ω	13.1388V	0.1752A
5	560Ω	6.2558V	0.0112A
6	820Ω	9.1603V	0.0112A
7	4.7Ω	0.9508V	0.2023A
8	1600Ω	0.9508V	0.0006A
9	22Ω	0.7953V	0.0362A
10	4.3Ω	0.1555V	0.0362A
11	2.7Ω	0.5118V	0.1895A
12	110Ω	20.8502V	0.1895A
13	430Ω	21.2828V	0.0495A
14	1.6Ω	0.0792V	0.0495A
Total	313.752Ω	75V	0.239A

Exercise #85

Find the voltage and amperage for each resistor.

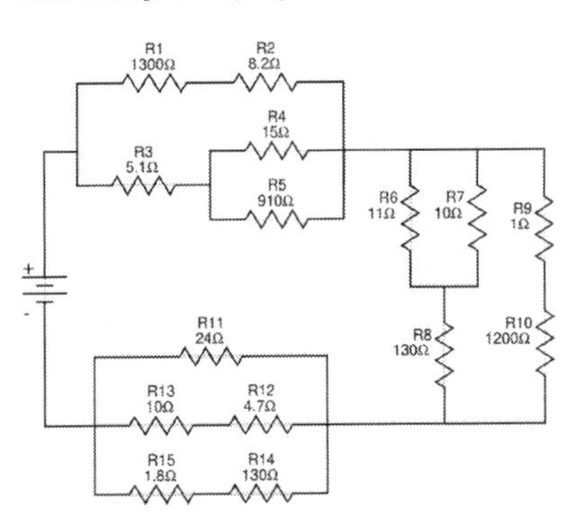

	R	V	I
1	1300Ω	31.175V	0.024A
2	8.2Ω	0.1966V	0.024A
3	5.1Ω	8.0575V	1.5799A
4	15Ω	23.3142V	1.5543A
5	910Ω	23.3142V	0.0256A
6	11Ω	7.551V	0.6865A
7	10Ω	7.551V	0.7551A
8	130Ω	187.4019V	1.4416A
9	1Ω	0.1623V	0.1623A
10	1200Ω	194.7905V	0.1623A
11	24Ω	13.6755V	0.5698A
12	4.7Ω	4.3724V	0.9303A
13	10Ω	9.3031V	0.9303A
14	130Ω	13.4887V	0.1038A
15	1.8Ω	0.1868V	0.1038A
Total	149.6373Ω	240V	1.6039A

Exercise #86

Find the voltage and amperage for each resistor.

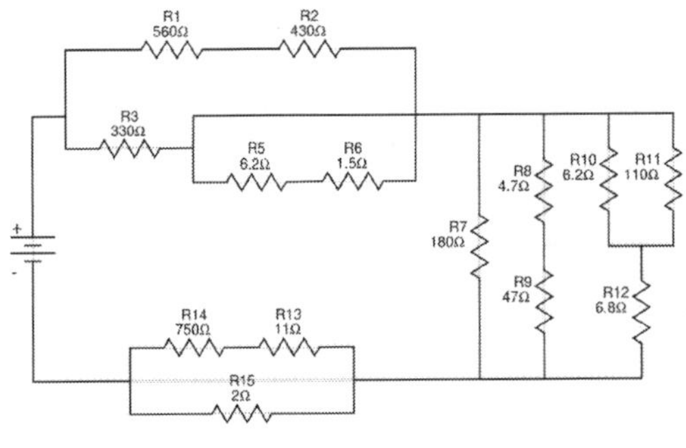

	R	V	I
1	560Ω	2.4326V	0.0043A
2	430Ω	1.8679V	0.0043A
3	330Ω	4.2275V	0.0128A
4	22Ω	0.0731V	0.0033A
5	6.2Ω	0.0588V	0.0095A
6	1.5Ω	0.0142V	0.0095A
7	180Ω	0.1652V	0.0009A
8	4.7Ω	0.015V	0.0032A
9	47Ω	0.1502V	0.0032A
10	6.2Ω	0.0765V	0.0123A
11	110Ω	0.0765V	0.0007A
12	6.8Ω	0.0887V	0.013A
13	11Ω	0.0005V	0A
14	750Ω	0.0337V	0A
15	2Ω	0.0342V	0.0171A
Total	262.3205Ω	4.5V	0.0172A

Exercise #87

Find the voltage and amperage for each resistor.

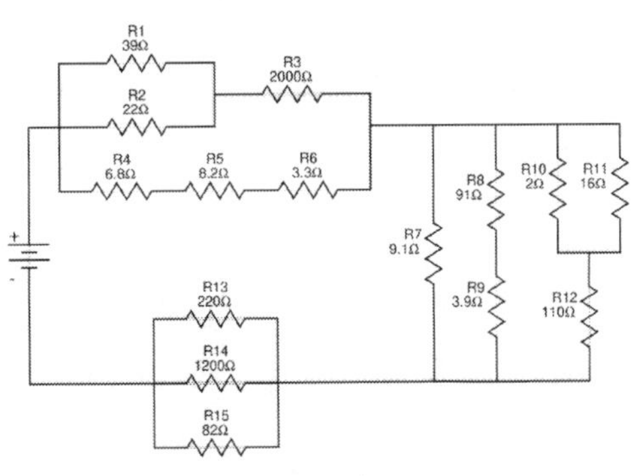

	R	V	I
1	39Ω	0.0459V	0.0012A
2	22Ω	0.0459V	0.0021A
3	2000Ω	6.5274V	0.0033A
4	6.8Ω	2.4425V	0.3592A
5	8.2Ω	2.9454V	0.3592A
6	3.3Ω	1.1854V	0.3592A
7	9.1Ω	2.8017V	0.3079A
8	91Ω	2.6865V	0.0295A
9	3.9Ω	0.1151V	0.0295A
10	2Ω	0.0446V	0.0223A
11	16Ω	0.0446V	0.0028A
12	110Ω	2.7571V	0.0251A
13	220Ω	20.625V	0.0938A
14	1200Ω	20.625V	0.0172A
15	82Ω	20.625V	0.2515A
Total	82.7673Ω	30V	0.3625A

Exercise #88

Find the voltage and amperage for each resistor.

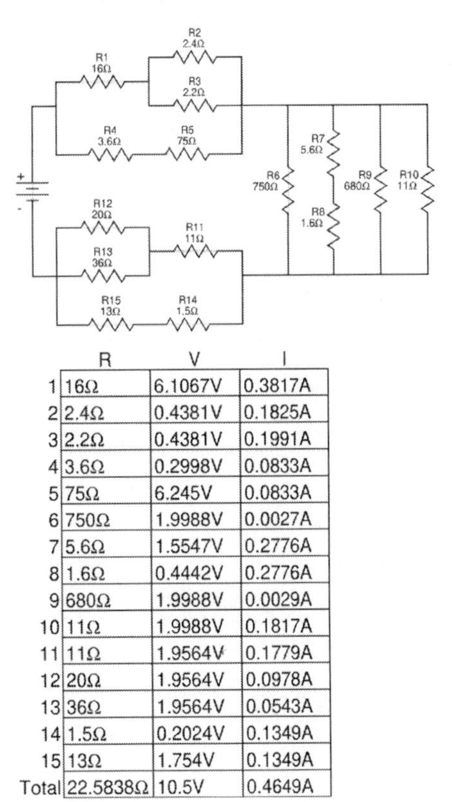

	R	V	I
1	16Ω	6.1067V	0.3817A
2	2.4Ω	0.4381V	0.1825A
3	2.2Ω	0.4381V	0.1991A
4	3.6Ω	0.2998V	0.0833A
5	75Ω	6.245V	0.0833A
6	750Ω	1.9988V	0.0027A
7	5.6Ω	1.5547V	0.2776A
8	1.6Ω	0.4442V	0.2776A
9	680Ω	1.9988V	0.0029A
10	11Ω	1.9988V	0.1817A
11	11Ω	1.9564V	0.1779A
12	20Ω	1.9564V	0.0978A
13	36Ω	1.9564V	0.0543A
14	1.5Ω	0.2024V	0.1349A
15	13Ω	1.754V	0.1349A
Total	22.5838Ω	10.5V	0.4649A

Exercise #89

Find the voltage and amperage for each resistor.

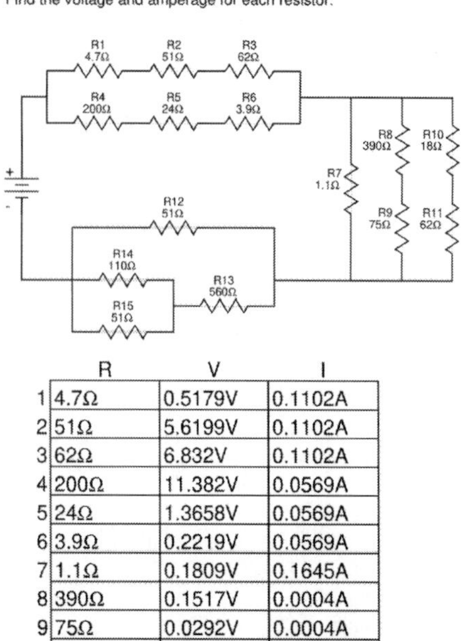

	R	V	I
1	4.7Ω	0.5179V	0.1102A
2	51Ω	5.6199V	0.1102A
3	62Ω	6.832V	0.1102A
4	200Ω	11.382V	0.0569A
5	24Ω	1.3658V	0.0569A
6	3.9Ω	0.2219V	0.0569A
7	1.1Ω	0.1809V	0.1645A
8	390Ω	0.1517V	0.0004A
9	75Ω	0.0292V	0.0004A
10	18Ω	0.0407V	0.0023A
11	62Ω	0.1402V	0.0023A
12	51Ω	7.8493V	0.1539A
13	560Ω	7.3895V	0.0132A
14	110Ω	0.4598V	0.0042A
15	51Ω	0.4598V	0.009A
Total	125.6705Ω	21V	0.1671A

Exercise #90

Find the voltage and amperage for each resistor.

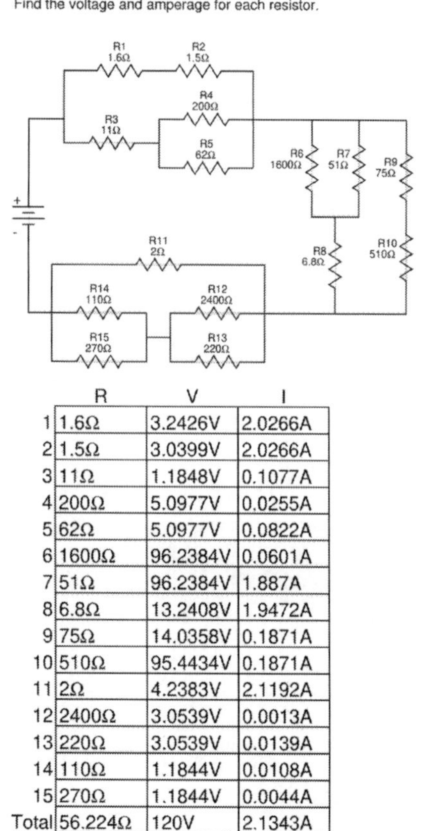

	R	V	I
1	1.6Ω	3.2426V	2.0266A
2	1.5Ω	3.0399V	2.0266A
3	11Ω	1.1848V	0.1077A
4	200Ω	5.0977V	0.0255A
5	62Ω	5.0977V	0.0822A
6	1600Ω	96.2384V	0.0601A
7	51Ω	96.2384V	1.887A
8	6.8Ω	13.2408V	1.9472A
9	75Ω	14.0358V	0.1871A
10	510Ω	95.4434V	0.1871A
11	2Ω	4.2383V	2.1192A
12	2400Ω	3.0539V	0.0013A
13	220Ω	3.0539V	0.0139A
14	110Ω	1.1844V	0.0108A
15	270Ω	1.1844V	0.0044A
Total	56.224Ω	120V	2.1343A

Exercise #91

Find the voltage and amperage for each resistor.

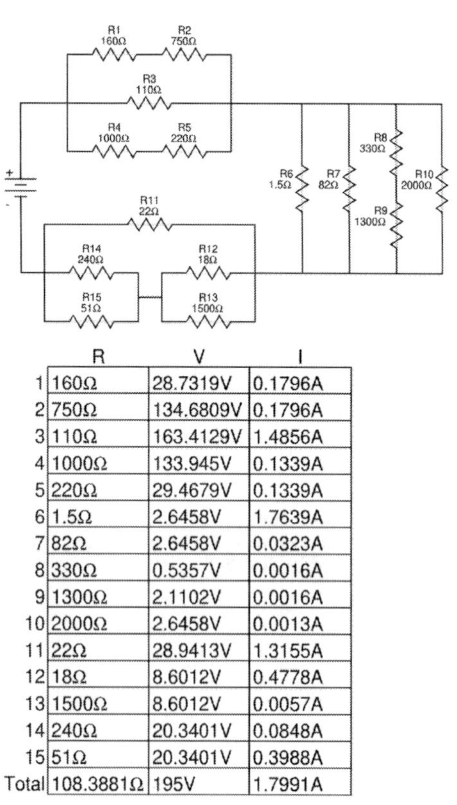

	R	V	I
1	160Ω	28.7319V	0.1796A
2	750Ω	134.6809V	0.1796A
3	110Ω	163.4129V	1.4856A
4	1000Ω	133.945V	0.1339A
5	220Ω	29.4679V	0.1339A
6	1.5Ω	2.6458V	1.7639A
7	82Ω	2.6458V	0.0323A
8	330Ω	0.5357V	0.0016A
9	1300Ω	2.1102V	0.0016A
10	2000Ω	2.6458V	0.0013A
11	22Ω	28.9413V	1.3155A
12	18Ω	8.6012V	0.4778A
13	1500Ω	8.6012V	0.0057A
14	240Ω	20.3401V	0.0848A
15	51Ω	20.3401V	0.3988A
Total	108.3881Ω	195V	1.7991A

Exercise #92

Find the voltage and amperage for each resistor.

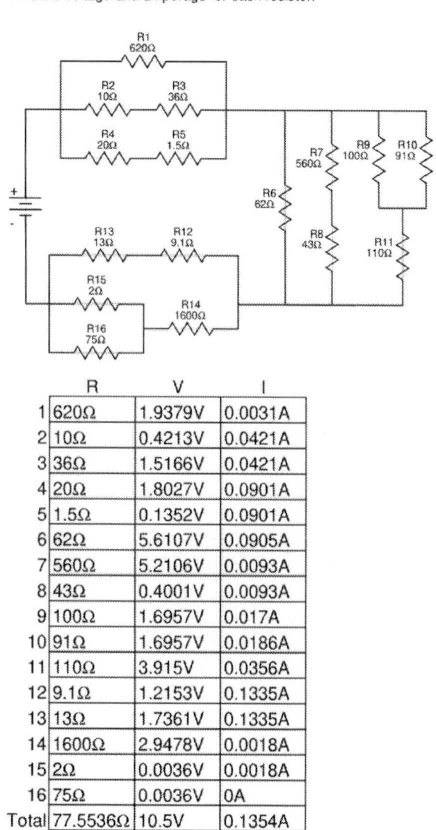

	R	V	I
1	620Ω	1.9379V	0.0031A
2	10Ω	0.4213V	0.0421A
3	36Ω	1.5166V	0.0421A
4	20Ω	1.8027V	0.0901A
5	1.5Ω	0.1352V	0.0901A
6	62Ω	5.6107V	0.0905A
7	560Ω	5.2106V	0.0093A
8	43Ω	0.4001V	0.0093A
9	100Ω	1.6957V	0.017A
10	91Ω	1.6957V	0.0186A
11	110Ω	3.915V	0.0356A
12	9.1Ω	1.2153V	0.1335A
13	13Ω	1.7361V	0.1335A
14	1600Ω	2.9478V	0.0018A
15	2Ω	0.0036V	0.0018A
16	75Ω	0.0036V	0A
Total	77.5536Ω	10.5V	0.1354A

Exercise #93

Find the voltage and amperage for each resistor.

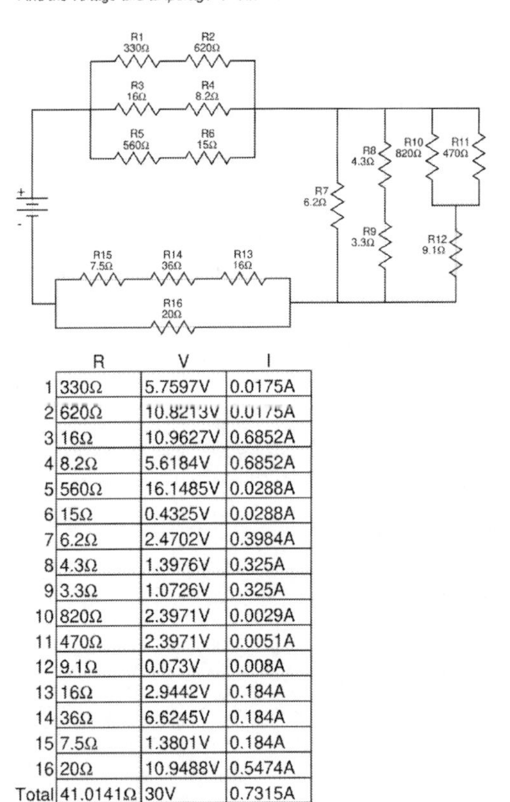

	R	V	I
1	330Ω	5.7597V	0.0175A
2	620Ω	10.8213V	0.0175A
3	16Ω	10.9627V	0.6852A
4	8.2Ω	5.6184V	0.6852A
5	560Ω	16.1485V	0.0288A
6	15Ω	0.4325V	0.0288A
7	6.2Ω	2.4702V	0.3984A
8	4.3Ω	1.3976V	0.325A
9	3.3Ω	1.0726V	0.325A
10	820Ω	2.3971V	0.0029A
11	470Ω	2.3971V	0.0051A
12	9.1Ω	0.073V	0.008A
13	16Ω	2.9442V	0.184A
14	36Ω	6.6245V	0.184A
15	7.5Ω	1.3801V	0.184A
16	20Ω	10.9488V	0.5474A
Total	41.0141Ω	30V	0.7315A

Exercise #94

Find the voltage and amperage for each resistor.

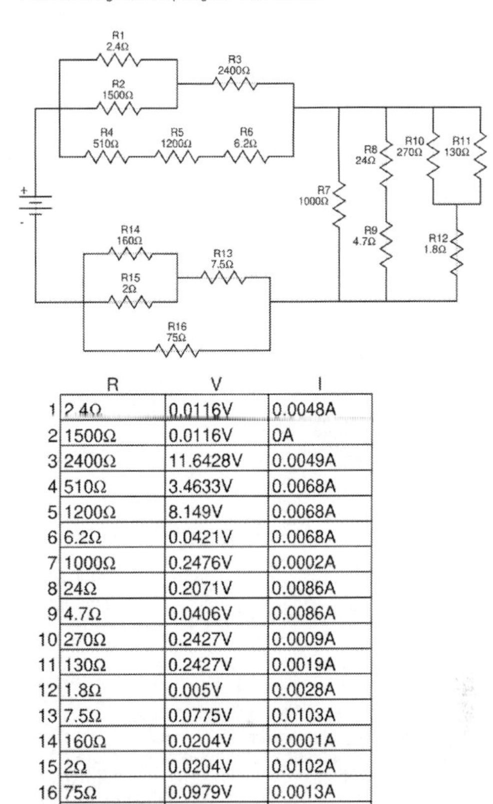

	R	V	I
1	2.4Ω	0.0116V	0.0048A
2	1500Ω	0.0116V	0A
3	2400Ω	11.6428V	0.0049A
4	510Ω	3.4633V	0.0068A
5	1200Ω	8.149V	0.0068A
6	6.2Ω	0.0421V	0.0068A
7	1000Ω	0.2476V	0.0002A
8	24Ω	0.2071V	0.0086A
9	4.7Ω	0.0406V	0.0086A
10	270Ω	0.2427V	0.0009A
11	130Ω	0.2427V	0.0019A
12	1.8Ω	0.005V	0.0028A
13	7.5Ω	0.0775V	0.0103A
14	160Ω	0.0204V	0.0001A
15	2Ω	0.0204V	0.0102A
16	75Ω	0.0979V	0.0013A
Total	1030.7519Ω	12V	0.0116A

Exercise #95

Find the voltage and amperage for each resistor.

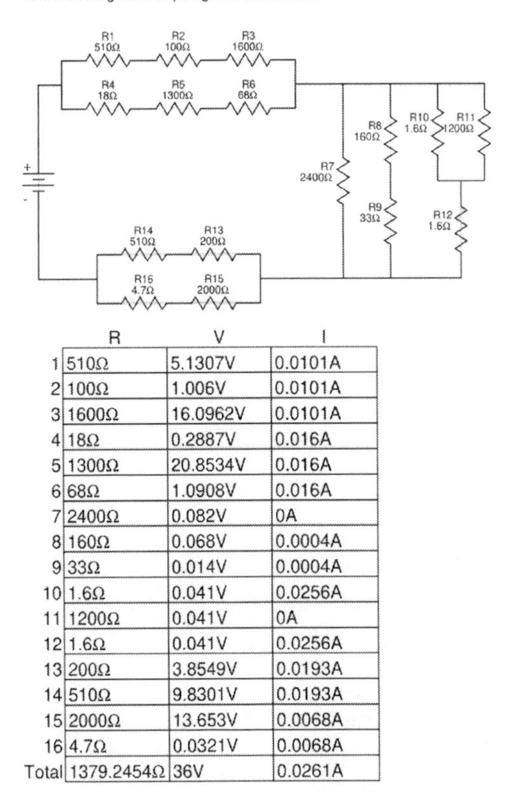

	R	V	I
1	510Ω	5.1307V	0.0101A
2	100Ω	1.006V	0.0101A
3	1600Ω	16.0962V	0.0101A
4	18Ω	0.2887V	0.016A
5	1300Ω	20.8534V	0.016A
6	68Ω	1.0908V	0.016A
7	2400Ω	0.082V	0A
8	160Ω	0.068V	0.0004A
9	33Ω	0.014V	0.0004A
10	1.6Ω	0.041V	0.0256A
11	1200Ω	0.041V	0A
12	1.6Ω	0.041V	0.0256A
13	200Ω	3.8549V	0.0193A
14	510Ω	9.8301V	0.0193A
15	2000Ω	13.653V	0.0068A
16	4.7Ω	0.0321V	0.0068A
Total	1379.2454Ω	36V	0.0261A

Exercise #96

Find the voltage and amperage for each resistor.

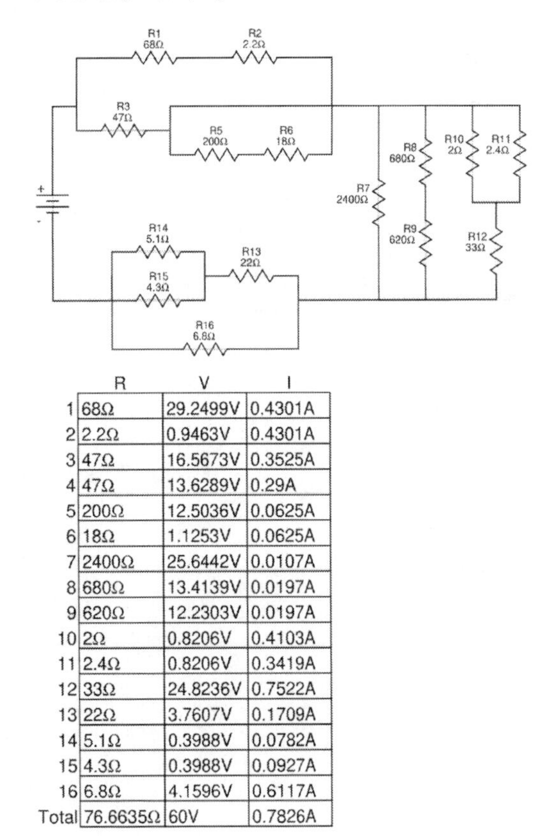

	R	V	I
1	68Ω	29.2499V	0.4301A
2	2.2Ω	0.9463V	0.4301A
3	47Ω	16.5673V	0.3525A
4	47Ω	13.6289V	0.29A
5	200Ω	12.5036V	0.0625A
6	18Ω	1.1253V	0.0625A
7	2400Ω	25.6442V	0.0107A
8	680Ω	13.4139V	0.0197A
9	620Ω	12.2303V	0.0197A
10	2Ω	0.8206V	0.4103A
11	2.4Ω	0.8206V	0.3419A
12	33Ω	24.8236V	0.7522A
13	22Ω	3.7607V	0.1709A
14	5.1Ω	0.3988V	0.0782A
15	4.3Ω	0.3988V	0.0927A
16	6.8Ω	4.1596V	0.6117A
Total	76.6635Ω	60V	0.7826A

Exercise #97

Find the voltage and amperage for each resistor.

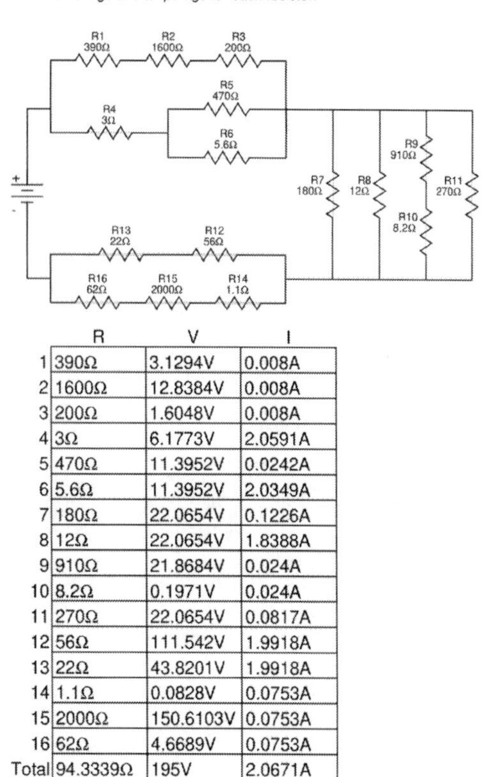

	R	V	I
1	390Ω	3.1294V	0.008A
2	1600Ω	12.8384V	0.008A
3	200Ω	1.6048V	0.008A
4	3Ω	6.1773V	2.0591A
5	470Ω	11.3952V	0.0242A
6	5.6Ω	11.3952V	2.0349A
7	180Ω	22.0654V	0.1226A
8	12Ω	22.0654V	1.8388A
9	910Ω	21.8684V	0.024A
10	8.2Ω	0.1971V	0.024A
11	270Ω	22.0654V	0.0817A
12	56Ω	111.542V	1.9918A
13	22Ω	43.8201V	1.9918A
14	1.1Ω	0.0828V	0.0753A
15	2000Ω	150.6103V	0.0753A
16	62Ω	4.6689V	0.0753A
Total	94.3339Ω	195V	2.0671A

Exercise #98

Find the voltage and amperage for each resistor.

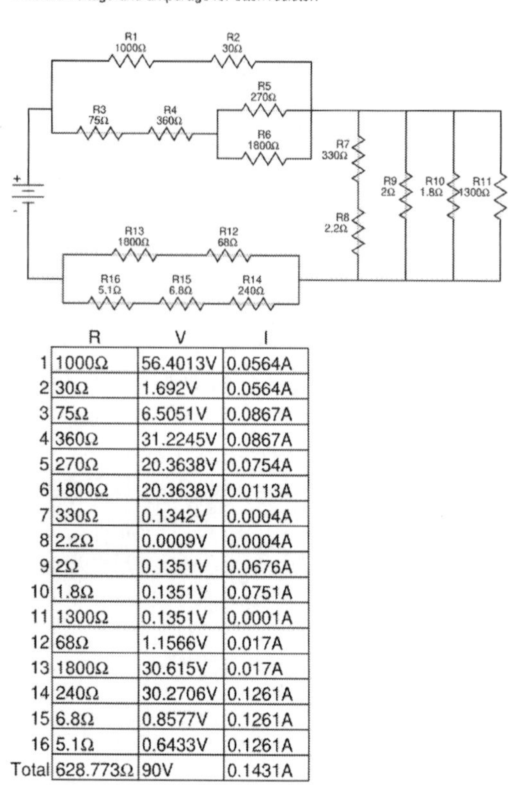

	R	V	I
1	1000Ω	56.4013V	0.0564A
2	30Ω	1.692V	0.0564A
3	75Ω	6.5051V	0.0867A
4	360Ω	31.2245V	0.0867A
5	270Ω	20.3638V	0.0754A
6	1800Ω	20.3638V	0.0113A
7	330Ω	0.1342V	0.0004A
8	2.2Ω	0.0009V	0.0004A
9	2Ω	0.1351V	0.0676A
10	1.8Ω	0.1351V	0.0751A
11	1300Ω	0.1351V	0.0001A
12	68Ω	1.1566V	0.017A
13	1800Ω	30.615V	0.017A
14	240Ω	30.2706V	0.1261A
15	6.8Ω	0.8577V	0.1261A
16	5.1Ω	0.6433V	0.1261A
Total	628.773Ω	90V	0.1431A

Exercise #99

Find the voltage and amperage for each resistor.

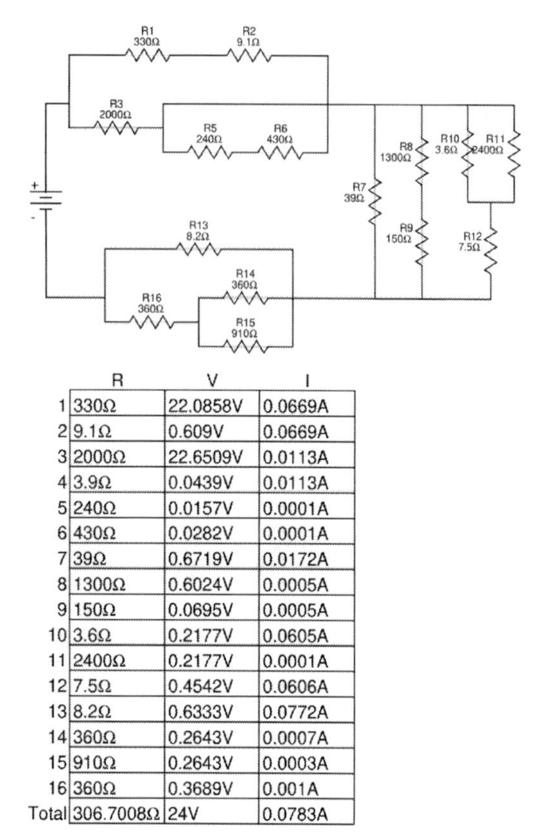

	R	V	I
1	330Ω	22.0858V	0.0669A
2	9.1Ω	0.609V	0.0669A
3	2000Ω	22.6509V	0.0113A
4	3.9Ω	0.0439V	0.0113A
5	240Ω	0.0157V	0.0001A
6	430Ω	0.0282V	0.0001A
7	39Ω	0.6719V	0.0172A
8	1300Ω	0.6024V	0.0005A
9	150Ω	0.0695V	0.0005A
10	3.6Ω	0.2177V	0.0605A
11	2400Ω	0.2177V	0.0001A
12	7.5Ω	0.4542V	0.0606A
13	8.2Ω	0.6333V	0.0772A
14	360Ω	0.2643V	0.0007A
15	910Ω	0.2643V	0.0003A
16	360Ω	0.3689V	0.001A
Total	306.7008Ω	24V	0.0783A

Exercise #100

Find the voltage and amperage for each resistor.

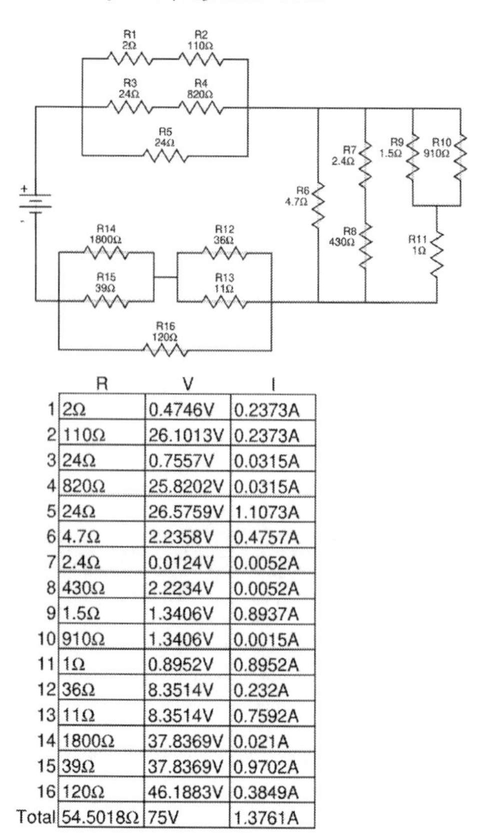

	R	V	I
1	2Ω	0.4746V	0.2373A
2	110Ω	26.1013V	0.2373A
3	24Ω	0.7557V	0.0315A
4	820Ω	25.8202V	0.0315A
5	24Ω	26.5759V	1.1073A
6	4.7Ω	2.2358V	0.4757A
7	2.4Ω	0.0124V	0.0052A
8	430Ω	2.2234V	0.0052A
9	1.5Ω	1.3406V	0.8937A
10	910Ω	1.3406V	0.0015A
11	1Ω	0.8952V	0.8952A
12	36Ω	8.3514V	0.232A
13	11Ω	8.3514V	0.7592A
14	1800Ω	37.8369V	0.021A
15	39Ω	37.8369V	0.9702A
16	120Ω	46.1883V	0.3849A
Total	54.5018Ω	75V	1.3761A

Made in the USA
Middletown, DE
30 August 2024

60002394R00071